Lecture Notes in Mathematics

Edited by A. Dold, Heic

T0222486

394

Walter Mead Patterson, 3rd
Major, United States Air Force
USAF Academy, Colorado, USA

Iterative Methods for the Solution of a Linear Operator Equation in Hilbert Space – A Survey

Springer-Verlag
Berlin · Heidelberg · New York 1974

AMS Subject Classifications (1970): Primary: 47-02, 47A50
Secondary: 46C10, 65J05

ISBN 3-540-06805-8 Springer-Verlag Berlin · Heidelberg · New York
ISBN 0-387-06805-8 Springer-Verlag New York · Heidelberg · Berlin

Offsetdruck: Julius Beltz, Hemsbach/Bergstr.

CONTENTS

CHAPTER I

INTRODUCTION

In this expository work we shall conduct a survey of iterative techniques for solving the linear operator equation $Ax=y$ in a Hilbert space. Whenever convenient these iterative schemes are given in the context of a complex Hilbert space. Chapter II is devoted to those methods (three in all) which are given only for real Hilbert space. Thus Chapter III covers those methods which are valid in a complex Hilbert space except for the two methods which are singled out for special attention in the last two chapters. Specifically, the method of successive approximations is covered in Chapter IV, and Chapter V consists of a discussion of gradient methods. While examining these techniques, our primary concern will be with the convergence of the sequence of approximate solutions. However, we shall often look at estimates of the error and at the speed of convergence of a method.

The Hilbert spaces which we consider here are infinite-dimensional but separable. In Chapter IV it will be more natural to present certain results in a Banach space context. We do not assume that any of the spaces discussed herein have any order properties. The operator A will always be linear but need not always be bounded. In the event that the operator A is unbounded, it will usually be assumed that the domain of A, denoted by $D(A)$, is dense in the Hilbert space under consideration. This condition will always be assumed to hold whenever mention is made of the adjoint operator A^*.

Our main interest will be in the case in which the equation $Ax=y$ has a solution. Therefore we shall not discuss such topics as generalized inverses of operators which arise when one considers an equation $Ax=y$ that might not be solvable even in a generalized sense. Usually conditions sufficient to insure the existence of a solution will be hypothesized. We shall not digress into a study of sufficient conditions for the existence of a solution.

Several other topics which will not be covered should be mentioned here. If one defines a function f on a Hilbert space $\mathcal{H}$ by $f(x) = Ax+x-y$, then x^* is a solution of $Ax=y$ if and only if x^* is a fixed point of f. Thus it would seem natural to consider fixed point theorems. However, a complete discussion of this topic is not possible here due to the vast scope of the field. A problem which is often closely associated with solving $Ax=y$ is the eigenvalue problem. Often there are closely related iterative schemes for producing eigenvalues of A. In spite of this, we do not discuss methods for finding eigenvalues. No attempt is made to discuss practical applications of the techniques considered. These methods are all theoretical in nature. At no time shall we examine the numerical or computational characteristics of our methods.

These notes are written for the reader who has had a basic course in functional analysis. Virtually any basic results used herein with which the reader might not be familiar are discussed in one or more of the works of Dunford and Schwartz [8], Taylor [41], or Riesz and Sz.-Nagy [38]. The notes have a single purpose. They should allow any qualified reader to obtain an overview of the wide range of techniques used in solving linear operator equations by iteration. Hopefully, a work such as this will facilitate the study of iterative methods by bringing together in an organized collection the numerous but widely scattered results now available in this field. In particular it is hoped that these notes will encourage persons working in the field of numerical analysis to apply their skills to the practical applications of these iterative methods which are treated only from a theoretical standpoint here.

Because of the expository nature of this paper, the ideas of many mathematicians are reproduced. All of these results are appropriately referenced. When the situation warranted, the statements of theorems have been changed by combining them, rearranging them, or placing them in a different context. On the other hand, many theorems are reproduced virtually without change. In some instances proofs have been improved and theorems have been generalized.

For the benefit of the reader who may not have extensive library facilities at hand, an attempt at completeness has been made. It is hoped that the reader will enjoy the leisurely pace of the exposition.

The author wishes to thank Professor Lawrence J. Lardy of Syracuse University for his valuable assistance during the writing of these notes. The author also wishes to thank both Syracuse University and the United States Air Force for their financial support of this project.

At this point we shall give some theorems and definitions which will be useful. It will always be assumed that $\mathcal{H}$ is a complex Hilbert space unless specific mention is made of it being real. First we define some types of operators in which we shall be interested.

Definition. (1)

A linear operator A in $\mathcal{H}$ is called positive if $\langle Au,u\rangle \geq 0$ for all $u \in D(A)$. The operator A is called positive definite if A is positive and $\langle Au,u\rangle = 0$ implies $u=0$. The operator A is called bounded below if there exists a real number m such that $\langle Au,u\rangle \geq m\|u\|^2$ for all $u \in D(A)$. In case $m=\gamma^2$ where $\gamma > 0$, we call A positive bounded below.

These definitions are stated here because each author employs a slightly different terminology. The reader is cautioned to check the exact meaning of each term when consulting other works. Now we state a basic lemma.

Lemma. (2)

If $\langle Au,u\rangle$ is real for all $u \in D(A)$, then A is symmetric. In particular, if A is positive, then A is symmetric.

The proof is obtained by using a polarization identity and the fact that $\langle Au,u\rangle = \overline{\langle u,Au\rangle} = \langle u,Au\rangle$ for $u \in D(A)$. Note that the proof fails and, in fact, the lemma is not true in a real Hilbert space. Indeed, the operator $A = \begin{pmatrix} 2 & -1 \\ 1 & 1 \end{pmatrix}$ in two-dimensional real Hilbert space is positive bounded below but not symmetric. Of course the operator A is not even positive, nor is $\langle Au,u\rangle$ even always real, if $\mathcal{H}$ is a two-dimensional complex Hilbert space.

In 1962 Petryshyn [116] introduced a class of operators which is more general than the positive bounded below operators. We shall often have occasion to use these operators when discussing results due to Petryshyn.

Definition. (3)

If $\mathcal{H}_1$ is a dense subspace of $\mathcal{H}$, then an operator K with domain $D(K) \supset \mathcal{H}_1$ is called continuously $\mathcal{H}_1$-invertible if the range of K, R(K), with K considered as an operator restricted to $\mathcal{H}_1$, is dense in $\mathcal{H}$ and K has a bounded inverse on R(K). An operator A is called K-positive definite (Kpd) if there exist a continuously D(A)-invertible closed operator K with $D(A) \subset D(K)$ and a constant $\alpha > 0$ such that

$$\langle Au,Ku\rangle \geq \alpha^2 \|Ku\|^2 \qquad (4)$$

for all $u \in D(A)$.

Really, in order to be consistent with Definition (1), we should call such operators K-positive bounded below, but we shall use Petryshyn's terminology, K-positive definite, here. We write Kpd for short. The choice $K=I$ shows that the positive bounded below operators are a subclass of the Kpd operators. If A is Kpd, then Petryshyn has shown [116] that in a complex Hilbert space $\langle Au,Kv\rangle = \langle Ku,Av\rangle$ for all $u, v \in D(A)$. This corresponds to Lemma (2).

For a densely defined, positive bounded below operator A , the domain of A can always be extended so that A will be self-adjoint and onto $\mathcal{H}$. This means that the equation $Ax=y$ will always have a unique solution for all $y \in \mathcal{H}$. The extension process is due to Friedrichs [16]. It involves defining an inner product by

$$\langle u,v\rangle_A = \langle Au,v\rangle \ , \quad u \ , \ v \in D(A) \ ,$$

and completing the resulting inner product space. In this Hilbert space, denoted by $F(A)$, we have

$$\|u\| \leq \frac{1}{\gamma} \|u\|_A \ , \quad u \in F(A) \ . \tag{5}$$

Petryshyn [116] has generalized the extension process of Friedrichs to Kpd operators. He defines an inner product by

$$\langle u,v\rangle_K = \langle Au,Kv\rangle \ , \quad u, v \in D(A) \ ,$$

and completes the resulting inner product space. In this Hilbert space, denoted by $F(A,K)$, we have

$$\|u\| \leq \frac{1}{\beta} \|u\|_K \ , \quad u \in F(A,K) \ . \tag{6}$$

The following theorem of Petryshyn generalizes the result referred to at the beginning of this paragraph.

<u>Theorem.</u> (7)

Every Kpd operator A with domain $D(A)$ dense in $\mathcal{H}$ can be extended to a closed Kpd operator A_e having a bounded inverse on $R(A_e) = \mathcal{H}$. The equation $A_e x=y$ has a unique solution x^* for every $y \in \mathcal{H}$.

In what follows, whenever A is Kpd, we shall assume that A has been extended and denote the extension A_e simply by A. A solution x^* of the equation $Ax=y$ which does not lie in the original domain of A is called a <u>generalized solution.</u>

To conclude we shall prove an easy lemma due to Mihlin [29] which is often useful.

<u>Lemma.</u> (8)

For a bounded operator A, if $Ax=y$ is solvable, then it is equivalent to the equation

$$A^*Ax = A^*y .$$

<u>Proof.</u>

If x_1 satisfies $Ax=y$, then obviously $A^*Ax_1 = A^*y$ also. If some x_2 also satisfies $A^*Ax = A^*y$, then $A^*A(x_2-x_1) = 0$ or $\langle A(x_2-x_1), A(x_2-x_1)\rangle = 0$. Thus $Ax_2 = Ax_1 = y$. ▌

CHAPTER II

ITERATIVE METHODS IN REAL HILBERT SPACES

SECTION 1

INTRODUCTION

In this chapter we shall examine some iterative schemes for solving the linear operator equation

$$Ax = y\ , \quad y \in \mathcal{Y}\ , \tag{1}$$

in a real Hilbert space. Certain of these schemes can be obtained as a special case of methods applicable in a complex Hilbert space, while others are variations of complex Hilbert space results which seem to hold only in a real Hilbert space. With the possible exception of the Samanskii method, the methods which we shall discuss in this chapter are quite simple in nature and, hence, form a good introduction to iterative methods in general. In Section 2 we shall examine a method due to Altman [51], but only after a discussion of several important inequalities which are required both here and in the sequel. In Section 3 we investigate another method of Altman [49] and a method of Samanskii [135]. The latter will play an important role in our discussion of gradient methods in Chapter V.

In a direct method such as the Ritz method or Galerkin method, one can compute the nth approximate solution x_n at once, without regard to $x_1, x_2, \ldots, x_{n-1}$. In an iterative method however, each approximate solution is a function of the preceding ones. In most instances, x_n can be computed from x_{n-1} alone, but, of course, x_{n-1} was computed from x_{n-2} and so forth. The initial approximation will always be denoted by x_0 so that x_n is the approximation computed at the $n^{\underline{th}}$ step. The procedure used in going from one step to the next is fixed in advance and is often given by a recursion formula. The recursion formula along with initial conditions and conditions on the operator A go together to make up an iterative scheme. One must note that the area of application of each method presented is limited. This may be due to restrictions placed on the operator, or may be due to more practical considerations such as speed of convergence.

As usual we shall denote by x^* the exact solution of (1) when it exists. The <u>error vector</u>, x^*-x , associated with a given $x \in \mathcal{X}$ will be denoted by $e(x)$. However, $e(x_n)$ will often be written e_n , and, when x is clear from the context, we may write e in place of $e(x)$. The <u>residual vector</u>, $y-Ax$, associated with a given $x \in \mathcal{X}$ will be denoted by $r(x)$, and the notational conventions concerning $e(x)$ apply to $r(x)$ as well. The relations $Ae=r$ and $A^{-1}r=e$ (when A^{-1} exists) are often used.

SECTION 2

BOUNDED, POSITIVE BOUNDED BELOW, AND SELF-ADJOINT OPERATORS

Using iterative methods, we wish to solve the linear operator equation

$$Ax = y , \quad y \in \mathcal{H} , \tag{1}$$

in a real Hilbert space $\mathcal{H}$, when A is bounded, positive bounded below, and self-adjoint. Let the constant $\gamma > 0$ be such that

$$\langle Ax,x\rangle \geq \gamma^2\|x\|^2$$

for all $x \in \mathcal{H}$. This condition shows that A must be 1-1 . Since A is self-adjoint and $D(A) = \mathcal{H}$, Theorem (I.1.7) shows that $R(A) = \mathcal{H}$. Hence (1) has a unique solution x^* for every $y \in \mathcal{H}$.

The first method that we wish to mention is due to Kolomý [86]. The following is a quotation of his theorem, which he states for a real Hilbert space $\mathcal{H}$.

Theorem. (2)

If A is a linear bounded operator in $\mathcal{H}$, and if there exists a positive number m such that for every $x \in \mathcal{H}$, the inequality

$$\langle Ax,x\rangle \geq m\|x\|^2$$

holds, then the sequence $\{x_n\}$ defined by the equations

$$x_n = \sum_{k=1}^{n} \beta_{k-1} y_{k-1} , \tag{3}$$

$$y_0 = y , \tag{4}$$

$$y_k = y_{k-1} - \beta_{k-1} A y_{k-1} , \tag{5}$$

$$\beta_k = \frac{\langle A y_k , y_k \rangle}{\|A y_k\|^2} , \tag{6}$$

converges in the norm of $\mathcal{H}$ to the solution $x^* \in \mathcal{H}$ of (1).

We do not prove this theorem here, since we shall prove a reformulation of it for a complex Hilbert space in Chapter III. In the complex case, it turns out that the result is a special case of a more general result due to Petryshyn [121] concerning operators forming an acute angle. This will be discussed in Chapter III. Therefore we leave Kolomý's result for the moment.

Before introducing our next method, we wish to state without proof an inequality established by Greub and Rheinboldt [17]. It is a generalization of an inequality due to Kantorovic [77]. From it we will obtain, as a corollary, an inequality that was first proved for finite-dimensional spaces by Krasnoselskii and Krein [27], which we shall require in the proof of the convergence of the next method. Also we shall be able to use our corollary to give a simple proof of a generalized inequality due to Petryshyn [116]. These results are valid in either real or complex Hilbert spaces;

the assumption that the operators are self-adjoint is superfluous in the complex case, but required in the real case. First, the theorem of Greub and Rheinboldt:

Theorem. (7)

Let A be a bounded, positive bounded below, self-adjoint operator on a Hilbert space $\mathcal{H}$. If m and M are respectively the greatest lower and least upper bounds of the spectrum of A, then

$$\langle Ax,x\rangle\langle A^{-1}x,x\rangle \leq \frac{(M+m)^2}{4Mm}\langle x,x\rangle^2 \tag{8}$$

for all $x \in \mathcal{H}$.

Corollary. (9)

With A, m, and M as in the theorem,

$$\inf_{\|z\|=1} \frac{\langle z,Az\rangle^2}{\|Az\|^2} \geq \frac{4Mm}{(M+m)^2} \tag{10}$$

and

$$\frac{\langle Au,Au\rangle\langle u,u\rangle}{\langle Au,u\rangle^2} \leq \frac{(M+m)^2}{4Mm}, \tag{11}$$

for all $u \in \mathcal{H}$, $u \neq 0$.

Proof.

Let $z = \frac{u}{\|u\|}$ where u is an arbitrary nonzero member of $\mathcal{H}$ so that $z \in \mathcal{H}$ is arbitrary with $\|z\| = 1$. Let $x \in \mathcal{H}$ be such that $x = A^{1/2}u$. Substituting in (8):

$$\frac{4Mm}{(M+m)^2} \le \frac{\langle x,x\rangle^2}{\langle Ax,x\rangle\langle A^{-1}x,x\rangle} = \frac{\langle A^{1/2}u,A^{1/2}u\rangle^2}{\langle A^{3/2}u,A^{1/2}u\rangle\langle A^{-1/2}u,A^{1/2}u\rangle}$$

$$= \frac{\langle Au,u\rangle^2}{\langle Au,Au\rangle\langle u,u\rangle} = \frac{\langle Az,z\rangle^2}{\|Az\|^2} ,$$

which implies both results. ▌

Altman [51] states that equality actually holds in (10). Both Greub and Rheinboldt [17] and Krasnoselskii and Krein [27] prove the equality in the finite-dimensional case. A proof of the equality in the general case is given in [144]. However, we shall only require the inequality here.

In order to prove the generalized inequality of Petryshyn, we require a lemma due to Reid [37]. It generalizes the well-known inequality $|\langle x,Ax\rangle| \le \|A\|\langle x,x\rangle$.

Lemma. (12)

If A and K are bounded linear operators on a Hilbert space $\mathcal{H}$ such that K is symmetric and positive and such that KA is symmetric, then

$$|\langle x,KAx\rangle| \le \|A\|\langle x,Kx\rangle \tag{13}$$

for all $x \in \mathcal{H}$.

Proof.

The symmetry of K and KA yields

$$\langle A^r x,KA^s x\rangle = \langle KA^{r+1}x,A^{s-1}x\rangle = \langle A^{r+1}x,KA^{s-1}x\rangle=\ldots=\langle KA^{r+s}x,x\rangle . \tag{14}$$

This shows that KA^n is symmetric for $n=0,1,2,\ldots$. Using the generalized Schwarz inequality along with the well-known $ab \le \frac{1}{2}(a^2+b^2)$ we have

$$|\langle x,Ky\rangle| \le \langle x,Kx\rangle^{1/2}\langle y,Ky\rangle^{1/2} \le \tfrac{1}{2}(\langle x,Kx\rangle + \langle y,Ky\rangle) .$$

Letting $y=A^r x$ and using (14), we obtain

$$\begin{aligned} |\langle x,KA^r x\rangle| &\le \tfrac{1}{2}\,(\langle x,Kx\rangle + \langle A^r x,KA^r x\rangle) \\ &= \tfrac{1}{2}\,(\langle x,Kx\rangle + \langle x,KA^{2r}x\rangle) \end{aligned}$$

for $r=1,2,\ldots$. By induction,

$$|\langle x,KAx\rangle| \le \left(\frac{1}{2}+\frac{1}{4}+\ldots+\frac{1}{2^r}\right)\langle x,Kx\rangle + \frac{1}{2^r}\langle x,KA^{2^r}x\rangle \tag{15}$$

for $r=1,2,\ldots$. If $\|A\| = 1$, we have for any p

$$|\langle x,KA^p x\rangle| \le \|K\|\;\|x\|^2 .$$

In this case the second term on the right in (15) goes to zero as $r \to \infty$. Thus $|\langle x,KAx\rangle| \le \langle x,Kx\rangle$ for $\|A\| = 1$. The result (13) is obvious if $\|A\| = 0$. If not, we use $A' = \frac{1}{\|A\|}A$ to obtain the result in general. ▌

Let A be a bounded and K*pd operator where K is a bounded and continuously invertible linear operator. Let B be a bounded linear operator and assume that there exists a constant $d > 0$ such that

$$\langle KAu,BAu\rangle \ge d\langle KAu,u\rangle . \tag{16}$$

Let $F(A,K^*)$ denote the generalization of the Friedrichs extension with the inner product

$$\langle u,v\rangle_{K^*} = \langle Au,K^*v\rangle = \langle KAu,v\rangle .$$

Note that since A is K^*pd, KA is positive bounded below. The following is due to Petryshyn [116].

Theorem. (17)

Let A be a bounded and K^*pd operator with K bounded and continuously invertible. Let the bounded operator B satisfy (16). Then BA is a bounded operator on $F(A,K^*)$, and if D denotes the norm of BA as an operator in $F(A,K^*)$, we have for all $u \neq 0$,

$$\frac{\langle KABAu,BAu\rangle\langle KAu,u\rangle}{\langle KAu,BAu\rangle^2} \leq \frac{(D+d)^2}{4Dd} . \qquad (18)$$

Proof.

Condition (16) shows that $\langle BAu,KAu\rangle \geq d\|u\|^2_{K^*}$. An imitation of the proof of Lemma (I.1.2) now yields

$$\langle KABAu,v\rangle = \langle BAu,KAv\rangle = \langle KAu,BAv\rangle = \langle u,KABAv\rangle .$$

Thus BA is a symmetric operator in $F(A,K^*)$. Now BA is a bounded operator on $\mathcal{H}$. Using Lemma (12):

$$\langle BAu,u\rangle_{K^*} = \langle KABAu,u\rangle \leq \|BA\|\langle KAu,u\rangle = \|BA\|\langle u,u\rangle_{K^*} .$$

Thus BA is also bounded as an operator on $F(A,K^*)$ or $D < +\infty$. Using (16) and the symmetry of BA in $F(A,K^*)$, we obtain $d\|u\|^2_{K^*} \leq \langle BAu,u\rangle_{K^*}$. Hence we see that $0 < dI \leq BA \leq DI$ in

$F(A,K^*)$. Now using inequality (11):

$$\frac{\langle KABAu,BAu\rangle\langle KAu,u\rangle}{\langle KAu,BAu\rangle^2} = \frac{\langle BAu,BAu\rangle_{K*}\langle u,u\rangle_{K*}}{\langle u,BAu\rangle_{K*}^2} \leq \frac{(D+d)^2}{4Dd} . \blacksquare$$

Replacing A by I , K by I , and B by A with the constant in the K*pd inequality at most one, we obtain (11) from (18) with $d=m$ and $D=M$. This shows that (11) is a special case of Petryshyn's result. Other choices for A, K, B, d, and D produce inequalities due to Polya-Szegö [36, p. 57] and Kantorovic [77, p. 106]. Petryshyn [116] shows that (11) and these two inequalities are, in fact, all equivalent. Thus his inequality can be viewed as a generalization of all three.

Now we turn our attention to a method due to Altman [51]. The method is of interest in its own right, but, in addition, it has a side result which is of interest from a computational standpoint. We shall assume that $0 < m \leq M < +\infty$ where m and M are the greatest lower and least upper bounds of the spectrum of A . Under these conditions if $y=0$ we know that $x^*=0$. Thus we assume without loss of generality that $\|y\| = 1$. Instead of (1), we consider the equation

$$Ax = \langle Ax,y\rangle y , \quad x \neq 0 , \quad \|y\| = 1 . \tag{19}$$

Notice that if $\bar{x}$ solves (19) then $x^* = \dfrac{\bar{x}}{\langle A\bar{x},y\rangle}$ solves (1), and if x^* solves (1) then it also solves (19). Since A is 1-1 , equation (19) shows that $\langle A\bar{x},y\rangle \neq 0$. We define a linear operator R by $Rx = Ax-\langle Ax,y\rangle y$, $\|y\| = 1$, and observe that this is the negative of the residual for equation (19).

<u>Theorem.</u> (20)

Let A be a bounded, positive bounded below, and self-adjoint operator on a real Hilbert space $\mathcal{H}$. Let x_0 be an arbitrary initial approximation, subject only to the condition

$$\langle x_0,y\rangle \neq 0 . \tag{21}$$

If the sequence $\{x_n\}$ is defined by

$$x_{n+1} = x_n - a_n Rx_n \tag{22}$$

with

$$a_n = \frac{\langle Rx_n, ARx_n\rangle}{\langle R^2x_n, R^2x_n\rangle} ,$$

then $x_n \to \bar{x}$ and $\dfrac{x_n}{\langle Ax_n,y\rangle} \to x^*$ at least as fast as a geometric progression with ratio $\frac{M-m}{M+m}$.

<u>Proof.</u>

Define $z_n=Rx_n$. By definition of the operator R , we have

$$\|Rx\|^2 = \|Ax\|^2 - \langle Ax,y\rangle^2 . \tag{23}$$

Also, by applying R to (22), we obtain $z_{n+1}= z_n-a_nRz_n$. Thus

$$\begin{aligned}\|z_{n+1}\|^2 &= \|z_n - a_n Rz_n\|^2 = \|z_n - a_n(Az_n - \langle Az_n, y\rangle y)\|^2 \\ &= \|z_n\|^2 - 2a_n(\langle Az_n, z_n\rangle - \langle Az_n, y\rangle\langle y, z_n\rangle) \\ &\quad + a_n^2(\|Az_n\|^2 - \langle Az_n, y\rangle^2) \ .\end{aligned}$$

But

$$\langle z_n, y\rangle = \langle Rx_n, y\rangle = \langle Ax_n, y\rangle - \langle Ax_n, y\rangle\langle y, y\rangle = 0 \tag{24}$$

for $n=0,1,2,\ldots$, so the use of (23) reduces the above to

$$\|z_{n+1}\|^2 = \|z_n\|^2 - a_n\langle z_n, Az_n\rangle \ . \tag{25}$$

This shows that $\{\|z_n\|\}$ is monotonically decreasing. We wish to show that this residual $z_n \to 0$. Noting that (25) can be written

$$\|z_{n+1}\|^2 = \|z_n\|^2 - \frac{\langle z_n, Az_n\rangle^2}{\|Rz_n\|^2} \ ,$$

we have

$$\frac{\|z_{n+1}\|^2}{\|z_n\|^2} = 1 - \frac{\langle z_n, Az_n\rangle^2}{\|z_n\|^2\|Rz_n\|^2} = 1 - \frac{\langle z_n, Az_n\rangle^2}{\|z_n\|^2\|Az_n\|^2} \cdot \frac{\|Az_n\|^2}{\|Rz_n\|^2} \ . \tag{26}$$

But

$$\frac{\|Az_n\|^2}{\|Rz_n\|^2} = \frac{\|Az_n\|^2 - \langle Az_n, y\rangle^2 + \langle Az_n, y\rangle^2}{\|Az_n\|^2 - \langle Az_n, y\rangle^2} = 1 + \frac{\langle Az_n, y\rangle^2}{\|Az_n\|^2 - \langle Az_n, y\rangle^2} \ .$$

Substitution in (26) yields

$$\|z_{n+1}\|^2 = \left\{1 - \frac{\langle z_n, Az_n\rangle^2}{\|z_n\|^2\|Az_n\|^2}\left(1 + \frac{\langle Az_n, y\rangle^2}{\|Az_n\|^2 - \langle Az_n, y\rangle^2}\right)\right\} \|z_n\|^2 \ .$$

Using inequality (11) with $q = \dfrac{4Mm}{(M+m)^2}$, we obtain

$$\|z_{n+1}\|^2 \le \left\{1 - q\left(1 + \frac{\langle Az_n,y\rangle^2}{\|Az_n\|^2 - \langle Az_n,y\rangle^2}\right)\right\} \|z_n\|^2$$

$$= \left\{1 - q\left(1 + \frac{\langle Az_n,y\rangle^2}{\|Rz_n\|^2}\right)\right\} \|z_n\|^2$$

$$\le (1-q)\|z_n\|^2 , \tag{27}$$

which shows that $z_n \to 0$. Hence $x_n \to \bar{x}$.

Finally we wish to show that

$$\inf\{|\langle Ax_n,y\rangle| : n=0,1,2,\ldots\} = c > 0 . \tag{28}$$

For any n, $\langle x_n,y\rangle = \langle x_0,y\rangle$ because $x_n = x_0 - \sum_{i=0}^{n-1} a_i z_i$ and $\langle z_i,y\rangle = 0$ for all i by (24). Thus

$$\inf\{\|x_n\| : n=0,1,2,\ldots\} > 0 . \tag{29}$$

If this were not so we would have a subsequence $x_{n_k} \to 0$ which would imply $\langle x_0,y\rangle = \langle x_{n_k},y\rangle \to 0$ contrary to condition (21). By (23) we have that

$$\|z_n\|^2 = \|Ax_n\|^2 - \langle Ax_n,y\rangle^2 \to 0 . \tag{30}$$

Since A is positive bounded below, (29) and (30) show that (28) holds. Hence $\dfrac{x_n}{\langle Ax_n,y\rangle} \to x^*$. Inequality (27) shows that convergence is at least as fast as that of a geometric progression with ratio

$$(1-q)^{1/2} = \frac{M-m}{M+m} . \blacksquare$$

Kantorovic's method of steepest descent [76, 77] also converges at least as fast as a geometric progression with ratio $\frac{M-m}{M+m}$. We shall discuss the method of steepest descent in Chapter V.

We can give a simple error estimate for Altman's method. Let c be as in (28). Then

$$\|z_n\| = \|Ax_n - \langle Ax_n, y\rangle y\| = |\langle Ax_n, y\rangle| \; \left\|A\left(\frac{x_n}{\langle Ax_n, y\rangle} - x^*\right)\right\|$$

$$\geq mc\left\|\frac{x_n}{\langle Ax_n, y\rangle} - x^*\right\|$$

so that

$$\left\|\frac{x_n}{\langle Ax_n, y\rangle} - x^*\right\| \leq \frac{\|z_n\|}{mc} . \tag{31}$$

In passing we wish to mention a side result connected with Altman's method. Notice in (27) that the speed of convergence for $\{z_n\}$ is dependent on $y \in \mathcal{X}$. One might expect that for certain y the convergence will be very rapid, while for others the convergence will be very slow. If the particular problem under consideration has in it a y for which convergence is "too slow", Altman [51] shows how one can introduce an arbitrary vector into the problem, which might speed up convergence. One merely tries arbitrary $v \in \mathcal{X}$ until a vector is found which yields a satisfactory rate of convergence. Briefly, Altman proceeds as follows:

Let $v \in \mathcal{H}$ be an arbitrary vector. If x^* solves (1), then $A(x^*+v) = y+Av$. Put

$$y' = \frac{y+Av}{\|y+Av\|} \quad \text{and} \quad x = \frac{x^*+v}{\|y+Av\|} .$$

Then $Ax=y'$. Just as (19) was obtained from (1), we obtain from this equation:

$$Ax = \langle Ax,y' \rangle y' \ , \quad x \neq 0 \ , \ \|y'\| = 1 . \tag{32}$$

Now if $\bar{x}$ solves (32), then

$$x^* = \frac{\|y+Av\|^2 \bar{x}}{\langle A\bar{x},y+Av \rangle} - v$$

solves (1). The theorem for this method is exactly parallel to Theorem (20). Condition (21) is replaced by $\langle x_0,y+Av \rangle \neq 0$. The sequence $\{x_n\}$ is given by

$$x_{n+1} = x_n - a_n R_1 x_n ,$$

where

$$R_1 x = Ax - \frac{\langle Ax,y+av \rangle}{\|y+Av\|^2} (y+Av) = Ax - \langle Ax,y' \rangle y'$$

and

$$z_n = R_1 x_n , \quad a_n = \frac{\langle z_n,Az_n \rangle}{\|R_1 z_n\|^2} .$$

If we let $c_1 = \inf\{ |\langle Ax_n,y+Av \rangle| : n=0,1,2,\dots\} > 0$, then the estimate (31) is replaced by

$$\left\| \left(\frac{\|y+Av\|^2}{\langle Ax_n, y+Av \rangle} x_n - v \right) - x^* \right\| \leq \frac{\|y+Av\|^2 \|z_n\|}{c_1 m} .$$

To conclude this section we mention two methods which are due to Bessmertnyh [53]. His recursion formula is of the form

$$x_{n+1} = x_n + \frac{c}{a_n + cb_n} r_n .$$

For specific choices of a_n and b_n, he shows the convergence of $\{x_n\}$ to the solution x^* of (1) for $c=1$ and $c=2$ in a real Hilbert space. This leads to speculation concerning the possible range of values for c which might produce convergence. These methods appear to be similar to the gradient type methods to be discussed in Chapter V. Since we shall prove the convergence of Bessmertnyh's methods in a complex Hilbert space in Chapter V (specifically for $c=2$), we do not discuss them further here.

SECTION 3

OTHER TYPES OF OPERATORS

In this section we shall again solve, by iterative means, the operator equation

$$Ax = y , \quad y \in \mathcal{H} , \tag{1}$$

in a real Hilbert space. The operator A will be bounded, just as in Section 2, but need not be positive bounded below or self-adjoint. However, we shall place some other restrictions on A. We present two methods, one due to Altman [49] and one due to Samanskii [135]. The proof of the method of Samanskii uses spectral theory.

The following theorem of Altman is an immediate corollary of a result which he proves in [49] for nonlinear operators. Since we do not wish to go into that result here, we shall give a direct proof of the theorem. The proof requires that $\mathcal{H}$ be a real Hilbert space. Petryshyn [116, p. 166] proves convergence in a complex Hilbert space for a method whose recursion formula differs from Altman's only slightly (see formula V. 2.18). In Petryshyn's formula there is no "2" in the denominator. It would thus seem that Altman's result might be modified to hold in a complex Hilbert space.

Theorem. (2)

Let A be a 1-1 bounded linear operator mapping a real

Hilbert space $\mathcal{H}$ onto itself satisfying

$$\|A\|\ \|A^{-1}\| \leq 2\alpha < 2 .$$

Then the sequence $\{x_n\}$ defined by

$$x_{n+1} = x_n + \frac{\|r_n\|^2}{2\|A^*r_n\|^2} A^*r_n \tag{3}$$

converges to the unique solution x^* of (1) for any choice x_0 of initial approximation.

Proof.

Using (3) we have

$$\langle x_{n+1}-x_n, 2A^*r_n\rangle = \frac{\|r_n\|^2}{2\|A^*r_n\|^2} \langle A^*r_n, 2A^*r_n\rangle = \|r_n\|^2 . \tag{4}$$

Thus

$$\begin{aligned} \|r_n\|^2 &= \|r_n\|^2 - \|r_{n-1}\|^2 + \langle x_n-x_{n-1}, 2A^*r_{n-1}\rangle \\ &= \|r_{n-1}-A(x_n-x_{n-1})\|^2 - \|r_{n-1}\|^2 + 2\langle A(x_n-x_{n-1}), r_{n-1}\rangle \\ &= \|r_{n-1}\|^2 - 2\langle r_{n-1}, A(x_n-x_{n-1})\rangle + \|A(x_n-x_{n-1})\|^2 \\ &\qquad - \|r_{n-1}\|^2 + 2\langle r_{n-1}, A(x_n-x_{n-1})\rangle \\ &= \|A(x_n-x_{n-1})\|^2 \\ &\leq \|A\|^2\|x_n-x_{n-1}\|^2 . \end{aligned} \tag{5}$$

Using (3) and then (5),

$$\|x_{n+1}-x_n\| = \frac{\|r_n\|^2}{2\|A^*r_n\|} = \frac{\|(A^*)^{-1}A^*r_n\|\ \|r_n\|}{2\|A^*r_n\|} \leq \frac{\|A^{-1}\|\ \|r_n\|}{2}$$

$$\leq \frac{\|A\|\ \|A^{-1}\|}{2} \|x_n-x_{n-1}\| \leq \alpha\|x_n-x_{n-1}\| . \qquad (6)$$

As a result:

$$\|x_{n+1}-x_n\| \leq \alpha^n\|x_1-x_0\| .$$

Thus $\{x_n\}$ must converge. Equation (5) shows that $r_n \to 0$ so that $x_n \to x^*$. ∎

The error may be estimated by

$$\|x_n-x^*\| \leq \frac{\|A^{-1}\|\ \|r_0\|}{2}\left(\frac{\alpha^n}{1-\alpha}\right) .$$

To see this, observe that

$$\|x_n-x^*\| = \|x_n-x_{n+1}+ x_{n+1}- x_{n+2}+ \dots - x^*\|$$

$$\leq \|x_n-x_{n+1}\| + \|x_{n+1}- x_{n+2}\| + \dots$$

$$\leq \alpha^n\|x_1-x_0\| + \alpha^{n+1}\|x_1-x_0\| + \dots$$

$$= \|x_1-x_0\|\left(\frac{\alpha^n}{1-\alpha}\right) .$$

But the first line of (6) shows that

$$\|x_1-x_0\| \leq \frac{\|A^{-1}\|\ \|r_0\|}{2} .$$

Hence the estimate.

The final method we shall discuss in this chapter is due to Samanskii [135]. Although the proof of this method is valid without change in a complex Hilbert space $\mathcal{H}$, the applications we shall make of the method require that $\mathcal{H}$ be real. Thus we shall present the theorem in that context. We shall require that the bounded operator A be such that R(A) is closed in the real Hilbert space $\mathcal{H}$. Such operators are often called normally solvable. (For example, see [40].) We notice at once that (1) will be solvable if and only if $y \in R(A)$; and, if such is the case, the solution will, in general, not be unique. Recall that for $y \in R(A)$, Lemma (I.1.8) shows that the equation

$$A^*Ax = A^*y \ , \quad y \in \mathcal{H} \ , \tag{7}$$

is equivalent to (1).

Consider the iteration process

$$x_{n+1} = q_n(B)x_0 + p_n(B)\tilde{y} \ , \tag{8}$$

where $B = A^*A$, q_n and p_n are polynomials, x_0 is an arbitrary initial approximation, and $\tilde{y}=A^*y$. It is desirable that the process be stationary. That is, if it should happen that x_n solves (1), then $x_n = x_{n+1}= x_{n+2}= \dots = x^*$. If we wish (8) to be stationary, we can achieve this if we put it in the form

$$x_{n+1} = x_0 - p_n(B)(Bx_0-\tilde{y}) \ . \tag{9}$$

For if $Ax_0=y$, then $Bx_0=\tilde{y}$ and $x_0=q_n(B)x_0 + p_n(B)\tilde{y}$ or $x_0 = [q_n(B) + p_n(B)B]x_0$. Thus it is natural to impose the restriction $I=q_n(B) + p_n(B)B$, which is a sufficient condition for (8) to be stationary. With $q_n(B) = I-p_n(B)B$, substitution in (8) yields (9).

If x^* solves (1), we have from (9) that

$$x_{n+1}-x^* = x_0 - p_n(B)(Bx_0-\tilde{y})-x^* + p_n(B)(Bx^*-\tilde{y})$$

$$= [I - p_n(B)B](x_0-x^*) \ . \qquad (10)$$

Denote the operator AA^* by $\tilde{B}$. Note that $A(A^*A)^m = (AA^*)^m A$, so that the operator $Ap_n(B)$ can be written as $p_n(\tilde{B})A$. We use this to obtain:

$$r_{n+1} = y - Ax_{n+1} = y - A[x_0 - p_n(B)(Bx_0-\tilde{y})]$$

$$= y - Ax_0 + Ap_n(B)(Bx_0-\tilde{y})$$

$$= r_0 + p_n(\tilde{B})AA^*(Ax_0-y)$$

$$= [I - p_n(\tilde{B})\tilde{B}]r_0 \ . \qquad (11)$$

If E_λ is the spectral resolution for B and if φ and Ψ are polynomials, then

$$\langle\varphi(B)x,\Psi(B)x\rangle = \int_{-\infty}^{\infty} \varphi(\lambda)\Psi(\lambda)d\langle E_\lambda x,x\rangle \ .$$

Thus from (10) and (11) we obtain:

$$\|x_{n+1}-x^*\|^2 = \int_{-\infty}^{\infty} [1 - p_n(\lambda)\lambda]^2 dg(\lambda) \ , \qquad (12)$$

where $g(\lambda) = \langle E_\lambda(x_0-x^*),x_0-x^*\rangle$, and

$$\|r_{n+1}\|^2 = \int_{-\infty}^{\infty} [1 - p_n(\lambda)\lambda]^2 dg_1(\lambda) \ ,$$

where $g_1(\lambda) = \langle\tilde{E}_\lambda r_0,r_0\rangle$. Here $\tilde{E}_\lambda$ is the spectral resolution for $\tilde{B}$. We note that B and $\tilde{B}$ are bounded and self-adjoint.

Theorem. (13)

Let A be a bounded linear operator with $R(A)$ closed, and let $B=A^*A$. Assume that $\sigma(B) \subset [m,M]$ where $0 \leq m \leq M$. If the sequence of polynomials $\{1-p_n(\lambda)\lambda\}$ is uniformly bounded on $[m,M]$ and converges to zero at each point of $[m,M]\setminus\{0\}$, then the sequence $\{x_n\}$ defined by (9) converges to a solution x^* of (1) provided $y \in R(A)$.

Proof.

If $m > 0$, then (12) and the Lebesgue dominated convergence theorem show that $\|x_{n+1}-x^*\| \to 0$. If $m=0$, we represent the nondecreasing function $g(\lambda)$ by $g(\lambda) = \mu(\lambda) + \eta(\lambda)$ where $\mu(\lambda)$ is a jump function and $\eta(\lambda)$ is continuous. Then

$$\int_{-\infty}^{\infty} [1-p_n(\lambda)\lambda]^2 dg(\lambda) = \sum_{k=1}^{\infty} [1-p_n(\lambda_k)\lambda_k]^2 \Delta_k + \int_0^M [1-p_n(\lambda)\lambda]^2 d\eta ,$$

where Δ_k is the jump of μ of λ_k. Now,

$$\sum_{k=1}^{\infty} [1-p_n(\lambda_k)\lambda_k]^2 \Delta_k \leq \sum_{k=1}^{\infty} c\Delta_k \leq c[g(M) - g(0)] < + \infty .$$

Thus the series converges and $\lim\limits_{n\to\infty} \sum\limits_{k=1}^{\infty} [1-p_n(\lambda_k)\lambda_k]^2 \Delta_k =$

$= \sum\limits_{k=1}^{\infty} \lim\limits_{n\to\infty} [1-p_n(\lambda_k)\lambda_k]^2 \Delta_k = 0$ provided $\lambda_k \neq 0$ for any k. In this case

$$\int_0^M [1-p_n(\lambda)\lambda]^2 d\eta = \int_0^{\epsilon} [1-p_n(\lambda)\lambda]^2 d\eta + \int_{\epsilon}^M [1-p_n(\lambda)\lambda]^2 d\eta ,$$

and the first integral can be made small (for any n) by making $\epsilon > 0$ small. Now for n sufficiently large, the second integral can also be made small. Thus

$$\lim_{n\to\infty} \int_0^M [1-p_n(\lambda)\lambda]^2 dg(\lambda) = 0$$

and $x_n \to x^*$.

Finally suppose that some $\lambda_k=0$, and note that λ_k is an eigenvalue of B. Write $x_0 = n_0 + m_0$ where $n_0 \in N(B)$ and $m_0 \in N(B)^\perp$. From (9):

$$x_{n+1} = n_0 + m_0 - p_n(B)(Bm_0-\tilde{y}) . \tag{14}$$

Observe that if $p_n(B) = \alpha_n B^n + \alpha_{n-1} B^{n-1} + \ldots + \alpha_1 B + \alpha_0$, where $\alpha_0 = p_n(0)$, then for any $n \in N(B)$,

$$\langle p_n(B)\tilde{y},n\rangle = \langle p_n(0)\tilde{y},n\rangle .$$

Now we can show that $x_{n+1}- n_0 \in N(B)^\perp$, since for any $n \in N(B)$,

$$\begin{aligned}\langle x_{n+1}-n_0,n\rangle &= \langle m_0,n\rangle - \langle p_n(B)Bm_0,n\rangle + \langle p_n(B)\tilde{y},n\rangle \\ &= \langle p_n(0)\tilde{y},n\rangle = p_n(0)\langle A^*y,n\rangle \\ &= p_n(0)\langle y,An\rangle .\end{aligned}$$

We have pointed out the equivalence of (1) and (7), so $Ax=0$ if and only if $Bx=A^*(0)=0$. But $Bn=0$, so $An=0$ and $\langle x_{n+1}-n_0,n\rangle = 0$. Thus we can write (14) as

$$x_{n+1}- n_0 = m_0- p_n(B)(Bm_0- \tilde{y})$$

and consider this as an iterative procedure on $N(B)^{\perp}$. If x^* solves (1) in $N(B)^{\perp}$, then $\{x_n - n_0\}$ converges to x^* as we showed above since zero is not eigenvalue of B on $N(B)^{\perp}$. ∎

Actually we have shown that $x_n \to n+m$, where n is the projection of x_0 onto $N(B)$ and m solves (1) for $y \in R(A)$.

In his applications of the method, Samanskii chooses the polynomials $p_n(\lambda)$ from the Chebyshev polynomials. For details, see [135].

A variation on the scheme (9) of Samanskii is given by Carusnikov [58]. He considers Samanskii's scheme (with $B=A$ self-adjoint) in the collection of Freidrichs spaces $F(A^{\alpha})$, α real, where the inner products are defined by

$$\langle u,v\rangle_{A^{\alpha}} = \langle A^{\alpha}u,v\rangle .$$

In this way he considers a sequence of polynomials $\{p_{n,\alpha}(A)\}$ for each real number α. We shall not consider the method in detail.

II.3

CHAPTER III

ITERATIVE METHODS IN COMPLEX HILBERT SPACES

SECTION 1

INTRODUCTION

In contrast with Chapter II, we shall now discuss iterative schemes for the solution of the linear operator equation

$$Ax = y , \quad y \in \mathcal{H} , \tag{1}$$

in a complex Hilbert space. All of the methods which we discuss here are also valid in a real Hilbert space provided, in the case in which the symmetry of a positive operator is used, one includes the symmetry of the operator as a hypothesis. Section 2 contains three very general methods due to Bialy [54]. In Chapter IV these methods will be related to a Banach space result of Petryshyn [118], which is also of interest in connection with the go-method to be discussed in Section 4.

In Section 3 we shall discuss two methods of Petryshyn [121] for the solution of an equation similar to (1). These methods apply to densely defined, unbounded operators of a quite general nature. Both methods use an auxiliary operator A for which equation (1) is easily and uniquely solvable. The methods consist, in part, of solving the simpler equation (1) repeatedly. This technique will appear again. The second method is applicable to operators which form an acute angle in the sense of Sobolevskii [141]. Instead of giving a proof of this method, we shall prove the theorem of Kolomý [86] which was mentioned in Section II.2. Then we shall show that this second method is a generalization of Kolomý's result and that the proof which we give for Kolomý's theorem can be easily modified to hold for the more general case.

Finally, in Section 4, we examine two methods which are generalizations of matrix methods to operator equations. The first method is due to Petryshyn [117, 119, 120, 126] and is called the go-method, a mnemonic for generalized overrelaxation iterative method. It generalizes the Gauss-Seidel and Jacobi methods of linear algebra. The second method is called the alternating direction method and is a generalization of the Peaceman-Rachford method of linear algebra. The formulation which we give is due to Kellogg [81].

SECTION 2

BIALY'S METHODS

In this section we shall discuss some very general iterative methods for solving the linear operator equation

$$Ax = y \ , \quad y \in \mathcal{H} \ , \tag{1}$$

where A is a bounded linear operator. These methods are due to Bialy [54]. They may be used even when (1) has no solution. By this we mean that the iterative schemes will produce a sequence $\{x_n\}$ such that

$$\|Ax_n - y\| \to \inf_{x \in \mathcal{H}} \|Ax - y\| \ .$$

Thus, in case (1) is solvable, we will have a solution, although not necessarily a unique one. We do not emphasize the case when (1) has no solution. Our main interest continues to be in an equation (1) which has a solution. In Section IV.1, Bialy's results are related to a Banach space technique of Petryshyn [118].

Let $N(A)$ denote the null space of A, and $M(A)$ its orthogonal complement. Denote by $P_N(A)$ and $P_M(A)$ the orthogonal projections of $\mathcal{H}$ onto $N(A)$ and $M(A)$ respectively. We shall prove two lemmas. The second lemma will be the key to proving convergence for all three of Bialy's methods.

<u>Lemma.</u> (2)

For every $y \in \mathcal{H}$,

$$\inf_{x\in\mathcal{H}} \|Ax-y\| = \|P_N(A^*)y\| . \tag{3}$$

Every solution $x' \in \mathcal{H}$ of $Ax=P_M(A^*)y$ satisfies

$$\inf_{x\in\mathcal{H}} \|Ax-y\| = \|Ax'-y\| . \tag{4}$$

Proof.

Since $\overline{R(A)} = M(A^*)$, we have $Ax \in M(A^*)$ for any $x \in \mathcal{H}$. Thus $Ax-P_m(A^*)y \in M(A^*)$. Writing $y = P_M(A^*)y + P_N(A^*)y$, we obtain

$$\|Ax-y\|^2 = \|Ax-P_M(A^*)y\|^2 + \|P_N(A^*)y\|^2$$

from the Pythagorean theorem. But $\inf_{x\in\mathcal{H}} \|Ax-P_M(A^*)y\| = 0$. Thus, if we take the infimum of the above expansion over all $x \in \mathcal{H}$, we obtain (3). If $Ax' = P_M(A^*)y$, then $Ax'-y = [P_M(A^*)-I]y = -P_N(A^*)y$ and we obtain (4). ▌

Lemma. (5)

Let A be a bounded and positive definite operator on $\mathcal{H}$. If α is such that $0 < \alpha < \frac{2}{\|A\|}$, then the sequence $\{u_n\}$ given by

$$u_{n+1} = u_n - \alpha A u_n$$

converges to zero for any choice of $u_0 \in \mathcal{H}$.

Proof.

Let $\{E_\lambda\}$ be the left resolution of the identity associated with A . Then, since zero is not an eigenvalue of A , the function

$$g(\lambda) = \sqrt{\langle E_\lambda u_0, u_0 \rangle}$$

is continuous at $\lambda=0$ with $g(0)=0$. Since $E_\lambda^2 = E_\lambda$, we have $g(\lambda) = \|E_\lambda u_0\|$. Thus $g(\|A\|+a) = \|u_0\|$ for all $a > 0$. The function $f(\lambda) = 1 - \alpha\lambda$ is monotonically decreasing. Since $0 < \alpha < \frac{2}{\|A\|}$, we can find an $a_0 > 0$ such that for all $\delta > 0$ with $\delta < \|A\| + a_0$, we have

$$q(\delta) = \max\{|f(\lambda)| : \delta < \lambda < \|A\| + a_0\} < 1 .$$

For every $\epsilon > 0$ there is a $\delta_0 > 0$ with $\delta_0 < \|A\| + a_0$ such that $g(\delta_0) = \|E_{\delta_0} u_0\| < \epsilon/2$. Also there exists a natural number n_0 such that

$$q^n(\delta_0)\|u_0 - E_{\delta_0} u_0\| < \epsilon/2 ,$$

for all $n > n_0$. Thus for $n > n_0$:

$$\|u_n\| = \|(I-\alpha A)^n u_0\| = \left\|\int_0^{\|A\|+a_0} (1-\alpha\lambda)^n \, dE_\lambda u_0\right\|$$

$$\leq \left\|\int_0^{\delta_0} (1-\alpha\lambda)^n \, dE_\lambda u_0\right\| + \left\|\int_{\delta_0}^{\|A\|+a_0} (1-\alpha\lambda)^n \, dE_\lambda u_0\right\|$$

$$\leq \left\|\int_0^{\delta_0} dE_\lambda u_0\right\| + q^n(\delta_0)\left\|\int_{\delta_0}^{\|A\|+a_0} dE_\lambda u_0\right\|$$

$$= \|E_{\delta_0} u_0\| + q^n(\delta_0)\|u_0 - E_{\delta_0} u_0\| < \frac{\epsilon}{2} + \frac{\epsilon}{2} = \epsilon . \blacksquare$$

When A is positive definite, the condition $\|I-\alpha A\| < 1$ implies that $0 < \alpha < \frac{2}{\|A\|}$ [38, p. 265]. If the converse were true, we would have $\|I-\alpha A\| < 1$ as a hypothesis for Lemma (5), and the lemma would be an obvious corollary to the basic method of successive approximations (Theorem(IV.1.6)). However, the condition $0 < \alpha < \frac{2}{\|A\|}$ need not imply that $\|I-\alpha A\| < 1$, but only that $\|I-\alpha A\| \leq 1$. (For example, consider the operator A on $L^2[0,1]$ given by $Ax(t) = tx(t)$. Choose $\alpha=1$. Since $\|A\| = 1$, clearly $0 < \alpha < \frac{2}{\|A\|}$. But $\|I-A\| = 1$.) The condition $0 < \alpha < \frac{2}{\|A\|}$ does imply that -1 is not an eigenvalue of $I-\alpha A$. Thus Lemma (5) can be obtained as a corollary to Krasnoselskii's Theorem (IV.2.1). However, we have chosen to give Bialy's original proof here.

Now that the key lemma is out of the way, we can prove convergence for the three methods of Bialy. Observe that if (1) has more than one solution, then, since the collection of all solutions is a closed convex set, there is a unique solution of smallest norm. We shall refer to such a solution as the minimal solution of (1), and denote it by $\hat{x}$.

Theorem. (6)

Let A be a bounded and positive operator on $\mathcal{H}$. If $x_0 \in \mathcal{H}$ and α is such that $0 < \alpha < \frac{2}{\|A\|}$, then for the sequence $\{x_n\}$ given by

$$x_{n+1} = x_n + \alpha r_n , \tag{7}$$

we have $Ax_n \to P_M(A)y$ and $\|Ax_n - y\| \to \inf_{x \in \mathcal{H}} \|Ax-y\|$. Further, $\{x_n\}$

converges if and only if (1) is solvable. If (1) is solvable, then $x_n \to P_N(A)x_0 + \hat{x}$.

Proof.

Writing r_n in the form $P_N(A)y + P_M(A)y - Ax_n$ and applying A to (7), we obtain

$$Ax_{n+1} = Ax_n + \alpha A(P_N(A)y + P_M(A)y - Ax_n) . \tag{8}$$

Since $AP_N(A)y=0$, we can put $v_{n+1}= Ax_{n+1}- P_M(A)y$ and rewrite (8) as

$$v_{n+1} = v_n - \alpha Av_n .$$

From the definition of v_{n+1} , we see that $v_{n+1} \in M(A)$ because $A=A^*$. Note that A is positive definite on $M(A)$. Thus we can apply Lemma (5) to obtain $v_n \to 0$. Hence $Ax_n \to P_M(A)y$, and Lemma (2) shows that $\|Ax_n-y\| \to \|P_N(A)y\| = \inf_{x\in\mathcal{X}} \|Ax-y\|$.

Now we apply $P_N(A)$ and $P_M(A)$ to (7). Noting that $P_N(A)Ax_n= 0$ and $P_M(A)Ax_n = AP_M(A)x_n$, we obtain

$$P_N(A)x_{n+1} = P_N(A)x_n + \alpha P_N(A)y = P_N(A)x_0 + (n+1)\alpha P_N(A)y \tag{9}$$

and

$$P_M(A)x_{n+1} = P_M(A)x_n - \alpha[AP_M(A)x_n - P_M(A)y] . \tag{10}$$

Assume that (1) is solvable. Then $y \in R(A)$, so $P_N(A)y=0$ and $P_M(A)y = y = A\hat{x}$. Putting $w_n = P_M(A)x_n - \hat{x}$, we have $w_n \in M(A)$. We rewrite (10) in the form

$$w_{n+1} = w_n - \alpha A w_n .$$

Using Lemma (5) again, we obtain $w_n \to 0$ or $P_M(A)x_n \to \hat{x}$. From (9), with $P_N(A)y=0$, we get $P_N(A)x_n = P_N(A)x_0$, so that

$$x_n = P_N(A)x_n + P_M(A)x_n \to P_N(A)x_0 + \hat{x} .$$

Now assume that (1) is not solvable or $y \notin R(A)$. If $P_N(A)y \neq 0$, then (9) shows that $\{x_n\}$ diverges. If $P_N(A)y=0$ and we assume that $x_n \to z$, then $Ax_n \to Az = P_M(A)y = y$ which contradicts $y \notin R(A)$.▌

It is of interest to note that this method bears a formal resemblance to the gradient methods which are discussed in Chapter V. The recursion formula (7) differs from a gradient method only in that α is fixed rather than being dependent on n . At the same time, we could write (7) in the form

$$x_{n+1} = (I-\alpha A)x_n + \alpha y .$$

This would make (7) a successive approximation method. Successive approximation methods are discussed in Chapter IV. It is this viewpoint which yields the close connection between Bialy's methods and a theorem of Petryshyn [118], which we shall discuss in Section IV.1.

The next method is suggestive of applying Theorem (6) to the equation $A^*Ax = A^*y$ which, as we have shown in Lemma (I.1.8), is equivalent to (1) when (1) is solvable.

III.2

Theorem. (11)

Let A be a bounded operator on $\mathcal{H}$. If $x_0 \in \mathcal{H}$ and α is such that $0 < \alpha < \frac{2}{\|A\|^2}$, then for the sequence $\{x_n\}$ given by

$$x_{n+1} = x_n + \alpha A^* r_n , \tag{12}$$

we have $Ax_n \to P_M(A^*)y$ and $\|Ax_n - y\| \to \inf_{x\in\mathcal{H}} \|Ax-y\|$. Further, $\{x_n\}$ converges if and only if

$$Ax = P_M(A^*)y \tag{13}$$

is solvable. If (13) is solvable, then $x_n \to P_N(A)x_0 + \tilde{x}$, where $\tilde{x}$ is the minimal solution of (13).

Proof.

Noting that $A^*P_N(A^*)y=0$, we apply A to (12) and obtain

$$Ax_{n+1} = Ax_n + \alpha AA^*[P_M(A^*)y - Ax_n] .$$

Setting $v_{n+1} = Ax_{n+1} - P_M(A^*)y \in M(A^*)$, we can write this in the form

$$v_{n+1} = v_n - \alpha AA^* v_n .$$

The operator AA^* is positive and positive definite on $M(AA^*)=M(A^*)$. Also $\|AA^*\| = \|A\|^2$. For $\|AA^*\| \le \|A\| \, \|A^*\| = \|A\|^2$, and $\|A^*x\|^2 = \langle A^*x, A^*x\rangle = \langle AA^*x, x\rangle \le \|AA^*\| \, \|x\|^2$ or $\|A\|^2 = \|A^*\|^2 \le \|AA^*\|$. Hence we may apply Lemma (5) with $0 < \alpha < \frac{2}{\|A\|^2} = \frac{2}{\|AA^*\|}$ on $M(A^*)$ to obtain $v_n \to 0$ or $Ax_n \to P_M(A^*)y$. Since $\|Ax_n - y\| \to \|P_N(A^*)y\|$, Lemma (2) shows that $\|Ax_n - y\| \to \inf_{x\in\mathcal{H}} \|Ax-y\|$.

Assume that $x_n \to z$. Then $Ax_n \to Az = P_M(A^*)y$ and (13) is solvable. Now assume that (13) is solvable, or $P_M(A^*)y \in R(A)$. Then, in particular, $A\tilde{x} = P_M(A^*)y$. Write (12) as

$$x_{n+1} = x_n - \alpha A^*A(x_n - \tilde{x}) . \tag{14}$$

Since $P_N(A)A^*x=0$ for all $x \in \mathscr{X}$, $P_N(A)x_{n+1} = P_N(A)x_n = \ldots = P_N(A)x_0$. Putting $w_n = P_M(A)x_n - \tilde{x} \in M(A)$, we rewrite (14) in the form

$$w_{n+1} = w_n - \alpha A^*Aw_n .$$

Since A^*A is positive definite on $M(A)$, we can use Lemma (5) to conclude $w_n \to 0$ or $P_M(A)x_n \to \tilde{x}$. Thus $x_n = P_N(A)x_n + P_M(A)x_n \to P_N(A)x_0 + \tilde{x}$. ▮

<u>Theorem.</u> (15)

Let A be a bounded and self-adjoint operator on $\mathscr{X}$. If $x_0 \in \mathscr{X}$ and α is such that $0 < \alpha < \frac{\sqrt{2}}{\|A\|}$, then for the sequence $\{x_n\}$ given by

$$x_{n+1} = x_n + (-1)^n \alpha r_n , \tag{16}$$

we have $Ax_n \to P_M(A)y$ and $\|Ax_n - y\| \to \inf_{x \in \mathscr{X}} \|Ax-y\|$. Further, $\{x_n\}$ converges if and only if (1) is solvable. If (1) is solvable, then $x_n \to P_N(A)x_0 + \hat{x}$.

<u>Proof</u>.

We first note that

$$\begin{aligned} x_{n+1} &= x_n + (-1)^n \alpha r_n \\ &= x_{n-1} + (-1)^{n-1} \alpha r_{n-1} + (-1)^n \alpha r_n \\ &= x_{n-1} - (-1)^{n-1} \alpha A x_{n-1} - (-1)^n \alpha A x_n \\ &= x_{n-1} - (-1)^{n-1} \alpha A x_{n-1} - (-1)^n \alpha [A x_{n-1} + (-1)^{n-1} \alpha A r_{n-1}] \\ &= x_{n-1} + \alpha^2 A r_{n-1} . \end{aligned}$$

Since $A^*=A$ and $0 < \alpha^2 < \frac{2}{\|A\|^2}$, this reduces to (12) with α^2 in place of α for the sequences $\{x_{2n}\}$ and $\{x_{2n+1}\}$, $n=0,1,2,\ldots$. We can apply Theorem (11) to obtain $Ax_{2n} \to P_M(A)y$, $Ax_{2n+1} \to P_M(A)y$, and $\|Ax_n-y\| \to \inf\limits_{x\in\mathscr{N}} \|Ax-y\|$. But (16) converges precisely when $\{x_{2n}\}$ and $\{x_{2n+1}\}$ converge to the same limit. If $P_M(A)y \in R(A)$ and $\tilde{x}$ is the minimal solution of (13), then Theorem (11) shows that

$$x_{2n} \to P_N(A)x_0 + \tilde{x} ,$$

and

$$x_{2n+1} \to P_N(A)x_1 + \tilde{x} = P_N(A)x_0 + \tilde{x} + \alpha P_N(A)y .$$

Thus $\{x_n\}$ converges if and only if, in addition to $P_M(A)y \in R(A)$, we have $P_N(A)y = 0$. But this means $y = P_M(A)y \in R(A)$. ▌

Bialy concludes his article [54] with two examples. One example is for a system of three equations in three unknowns, while the other concerns an integral equation. We do not reproduce these examples here.

III.2

SECTION 3

TWO METHODS OF PETRYSHYN

In this section we shall be concerned with two operator equations. Our goal will be to solve the operator equation

$$Lx = y' \ , \quad y' \in \mathcal{N} \ , \tag{1}$$

where L is an arbitrary unbounded operator. However, we shall also consider the equation

$$Ax = y \ , \quad y \in \mathcal{N} \ , \tag{2}$$

where the operator A is such that (2) is easily solvable, at least for certain $y \in \mathcal{N}$. The problem of solving (1) is reduced to that of repeatedly solving (2). Two methods are presented; both are due to Petryshyn [121].

Before giving Petryshyn's second method, we shall prove the theorem of Kolomý, Theorem (II.2.2), in a complex Hilbert space. Our formulation of that theorem will be different from that given earlier. We then show that Kolomý's theorem is a special case of Petryshyn's second method. Instead of giving the proof of Petryshyn's second method, we shall merely indicate how our proof of Kolomý's result can be generalized.

Let A be any Kpd operator. From Definition (I.1.3) we know that we can find $\alpha > 0$ such that

$$\langle Au, Ku \rangle \geq \alpha^2 \|Ku\|^2 \ , \quad u \in D(A) \ .$$

The definition requires that K be continuously D(A)-invertible. Hence, we can find a $\beta > 0$ such that $\|Ku\| \geq \beta\|u\|$ for $u \in D(A) \subset D(K)$. By letting $\alpha_1 = \alpha^2\beta^2 > 0$, we obtain

$$\langle Au,Ku\rangle \geq \alpha_1\|u\|^2 , \quad u \in D(A) . \tag{3}$$

We observe that just as in Lemma (I.1.2) one has $\langle Au,Kv\rangle = \langle Ku,Av\rangle$ for all $u, v \in D(A)$. (When $\mathcal{H}$ is a real Hilbert space this property must be assumed.) In this section we use the inner product

$$\langle u,v\rangle_K = \langle Au,Kv\rangle , \quad u, v \in D(A) ,$$

and the corresponding norm $\|\cdot\|_K$. The space $F(A,K) \subset \mathcal{H}$ is formed as discussed in Chapter I. The operator A will be assumed to have been extended as in Theorem (I.1.7).

Now assume that we are given an operator L which is densely defined in $\mathcal{H}$. We choose a Kpd operator A so that $D(A) = D(L)$ and such that for constants $\eta_1 > 0$ and $\eta_2 > 0$,

$$\mathrm{Re}\langle Lu,Ku\rangle \geq \eta_1\langle Au,Ku\rangle , \quad u \in D(L) , \tag{4}$$

and

$$|\langle Lu,Kv\rangle|^2 \leq \eta_2^2\langle Au,Ku\rangle\langle Av,Kv\rangle \tag{5}$$

for all $u, v \in D(L)$.

Define an operator $W = A^{-1}L$ on $D(L)$. As an operator in $F(A,K)$, W is bounded. For by (5):

$$\|Wu\|_K^4 = \langle Wu,Wu\rangle_K^2 = \langle Lu,K(A^{-1}Lu)\rangle^2$$
$$\leq \eta_2^2 \langle Au,Ku\rangle\langle Lu,KA^{-1}Lu\rangle$$
$$= \eta_2^2 \|u\|_K^2 \|Wu\|_K^2 .$$

Thus

$$\|Wu\|_K \leq \eta_2 \|u\|_K . \tag{6}$$

Let $\overline{W}$ denote the bounded closure of W in $F(A,K)$. Then, in view of the continuity of the inner product $\langle\cdot,\cdot\rangle_K$, using (4):

$$\operatorname{Re}\langle\overline{W}u,u\rangle_K = \operatorname{Re}\langle Lu,Ku\rangle \geq \eta_1\langle Au,Ku\rangle = \eta_1\|u\|_K^2 \tag{7}$$

holds for all $u \in F(A,K)$. Thus $\overline{W}$ is 1-1, or $\overline{W}$ has an inverse $\overline{W}^{-1}$ which is defined on $R(\overline{W}) = F(A,K)$. Using (7), we have $\|\overline{W}u\|_K\|u\|_K \geq |\langle\overline{W}u,u\rangle_K| \geq \operatorname{Re}\langle\overline{W}u,u\rangle_K \geq \eta_1\|u\|_K^2$, or

$$\|\overline{W}u\|_K \geq \eta_1\|u\|_K , \quad u \in F(A,K) . \tag{8}$$

Thus $\overline{W}^{-1}$ is bounded on $F(A,K)$. In [117] Petryshyn shows that L satisfying (4) and (5) can be extended to a closed operator which is continuously invertible. Thus (1) will have, perhaps in the generalized sense, a unique solution x^* for every $y' \in \mathcal{Y}$.

Before we can prove the first theorem of Petryshyn, we require a technical lemma.

<u>Lemma.</u> (9)

For the real number k given by

$$k = \inf\left\{\frac{\mathrm{Re}\langle \overline{W}^{-1}u,u\rangle_K}{\langle u,u\rangle_K} : u \in F(A,K) \;, \; u \neq 0\right\}$$

we have

$$k \geq \frac{\eta_1}{\eta_2^2} . \tag{10}$$

Proof.

Let $v = \overline{W}u$ in (7) to obtain:

$$\frac{\mathrm{Re}\langle v,\overline{W}^{-1}v\rangle_K}{\langle v,v\rangle_K} \geq \frac{\eta_1 \|\overline{W}^{-1}v\|_K^2}{\langle v,v\rangle_K} .$$

From (6) we have $\|\overline{W}\|_K \leq \eta_2$. Thus,

$$k = \inf\left\{\frac{\mathrm{Re}\langle \overline{W}^{-1}v,v\rangle_K}{\langle v,v\rangle_K} : v \in F(A,K),\; v \neq 0\right\}$$

$$\geq \eta_1 \inf\left\{\frac{\|\overline{W}^{-1}v\|_K^2}{\|v\|_K^2} : v \in F(A,K),\; v \neq 0\right\} = \frac{\eta_1}{\|\overline{W}\|_K^2} \geq \frac{\eta_1}{\eta_2^2} . \blacksquare$$

Theorem. (11)

Let the operators A, K, L, and W be as given above. If γ_1 is a real number such that

$$0 < \gamma_1 < 2k\eta_1 , \tag{12}$$

if $x_0 \in D(A)$ is an arbitrary initial approximation to the solution x^* of (1), and if $Az_0 = Lx_0 - y'$, then the sequence $\{x_n\}$ determined by

$$x_{n+1} = x_n - \gamma_1 t_n z_n , \quad n=0,1,2,\ldots , \tag{13}$$

where

$$t_n = \frac{\langle Az_n, Kz_n\rangle}{\mathrm{Re}\langle Lz_n, Kz_n\rangle} , \quad n=0,1,2,\ldots , \tag{14}$$

and

$$Az_{n+1} = Az_n - \gamma_1 t_n Lz_n , \quad n=0,1,2,\ldots , \tag{15}$$

converges, monotonically in the $\|\cdot\|_K$ norm, to the unique solution x^* of (1).

Proof.

Denoting the error vector by e_i as usual, we let $w_i = -e_i = x_i - x^*$. Then

$$w_{n+1} = w_n - \gamma_1 t_n z_n , \quad n=0,1,2,\ldots . \tag{16}$$

Note that $Az_0 = Lx_0 - y' = L(x_0 - x^*) = Lw_0$, and $Az_1 = Az_0 - \gamma_1 t_0 Lz_0 = L(x_0 - \gamma_1 t_0 z_0) - Lx^* = L(x_1 - x^*) = Lw_1$. If we assume that $Az_i = Lw_i$ for $i=1,2,\ldots,n$, then

$$\begin{aligned} Az_{n+1} &= Az_n - \gamma_1 t_n Lz_n = Lw_n - \gamma_1 t_n Lz_n \\ &= L(x_n - x^*) - \gamma_1 t_n Lz_n = L(x_n - \gamma_1 t_n z_n) - Lx^* \\ &= L(x_{n+1} - x^*) = Lw_{n+1} . \end{aligned}$$

Hence, by induction,

$$Az_n = Lw_n = Lx_n - y' , \quad n=0,1,2,\ldots . \tag{17}$$

From (16) we obtain

$$\langle Aw_{n+1}, Kw_{n+1}\rangle = \|w_{n+1}\|_K^2$$

$$= \|w_n\|_K^2 - 2\gamma_1 t_n \operatorname{Re}\langle z_n, w_n\rangle_K + \gamma_1^2 t_n^2 \|z_n\|_K^2 . \tag{18}$$

Now (17) may be written in the form $\overline{W}w_n = z_n$ or $w_n = \overline{W}^{-1} z_n$, $n=0,1,2,\ldots$. Thus, using (14), we may write (18) as

$$\|w_{n+1}\|_K^2 = \|w_n\|_K^2 - 2\gamma_1 \frac{\langle z_n, z_n\rangle_K^3}{(\operatorname{Re}\langle \overline{W}z_n, z_n\rangle_K)^2}\Big\{-\frac{1}{2}\gamma_1 + \frac{\operatorname{Re}\langle \overline{W}z_n, z_n\rangle_K \operatorname{Re}\langle z_n, \overline{W}^{-1}z_n\rangle_K}{\langle z_n, z_n\rangle_K^2}\Big\} . \tag{19}$$

Using the inequalities (7) and (10), we can express (19) as

$$\|w_{n+1}\|_K^2 \leq \|w_n\|_K^2 - \gamma_1(2k\eta_1 - \gamma_1)\frac{\langle z_n, z_n\rangle_K^3}{(\operatorname{Re}\langle \overline{W}z_n, z_n\rangle_K)^2} \tag{20}$$

because

$$\frac{\operatorname{Re}\langle \overline{W}z_n, z_n\rangle_K \operatorname{Re}\langle z_n, \overline{W}^{-1}z_n\rangle_K}{\langle z_n, z_n\rangle_K^2} \geq k\eta_1 .$$

Inequality (12) shows that $2k\eta_1 - \gamma_1 > 0$. Thus the sequence $\{\|w_n\|_K\}$ is monotonically decreasing. Hence it converges to some nonnegative real number. This means that

$$\lim_{n\to\infty} \frac{\langle z_n, z_n\rangle_K^3}{(\operatorname{Re}\langle \overline{W}z_n, z_n\rangle_K)^2} = \lim_{n\to\infty} \frac{\langle Az_n, Kz_n\rangle^3}{(\operatorname{Re}\langle Lz_n, Kz_n\rangle)^2} = 0 .$$

Using inequalities (5) and (8), we see that

$$\frac{\langle Az_n,Kz_n\rangle^3}{(\mathrm{Re}\langle Lz_n,Kz_n\rangle)^2} \geq \frac{1}{\eta_2^2}\langle Az_n,Kz_n\rangle = \frac{1}{\eta_2^2}\|z_n\|_K^2$$

$$= \frac{1}{\eta_2^2}\|\overline{W}w_n\|_K^2 \geq \left(\frac{\eta_1}{\eta_2}\right)^2 \|w_n\|_K^2 .$$

This implies that $w_n \to 0$ in $F(A,K)$. ▌

Notice how, instead of solving (1), we solve an equation of the form (2), namely (15), at each iteration. Inequality (3) shows that $\|u\|_K \geq \alpha_1^{1/2}\|u\|$ for all $u \in D(A)$, so that $x_n \to x^*$ in $\mathscr{H}$ as well as in $F(A,K)$.

Petryshyn also shows that if $D(K) = D(A)$, then we have $Lx_n \to y'$ in $\mathscr{H}$. Under this assumption he derives the following error estimate:

$$\|x_n - x^*\|_K \leq \frac{1}{\alpha\eta_1}\|Lx_n - y'\| .$$

The parabola given by $f(x) = x(2t-x)$ has its maximum when $x=t$. Examination of (20) thus reveals that a good choice of γ_1 should be $\gamma_1 = k\eta_1$. For this choice we obtain the most rapid rate of decrease for $\|w_n\|_K$. More specifically, we then have

$$\|w_{n+1}\|_K \leq q\|w_n\|_K , \quad q = \left[1 - \left(\frac{\eta_1}{\eta_2}\right)^6\right]^{1/2} < 1 ,$$

so that convergence is at least as fast as a geometric progression with ratio q . To see this, we use Lemma (9), $\|z_n\|_K^2 \geq \eta_1^2\|w_n\|_K^2$

from (8), and $\frac{\eta_1}{\|W\|_K^2} \geq \frac{\eta_1}{\eta_2^2}$ from (6). Then:

$$\|w_{n+1}\|_K^2 \leq \|w_n\|_K^2 - k^2 \eta_1^2 \frac{\langle z_n, z_n\rangle_K^3}{(\mathrm{Re}\langle Wz_n, z_n\rangle_K)^2}$$

$$\leq \|w_n\|_K^2 - k^2\eta_1\left(\frac{\eta_1}{\|W\|_K^2}\right)\left(\frac{\|z_n\|_K^6}{\|z_n\|_K^4}\right)$$

$$\leq \|w_n\|_K^2 - \left(\frac{\eta_1}{\eta_2^2}\right)^2\left(\frac{\eta_1^2}{\eta_2^2}\right)\|z_n\|_K^2$$

$$\leq \|w_n\|_K^2 - \left(\frac{\eta_1^6}{\eta_2^6}\right)\|w_n\|_K^2 = q^2\|w_n\|_K^2 .$$

Now we wish to prove the theorem of Kolomý [86] given earlier as Theorem (II.2.2). With regard to Kolomý's recursion formulas (II.2.3) through (II.2.6), one can easily show that $x_{n+1} = x_n + \beta_n y_n$. Also one can show that y_n is actually the residual r_n. For if $x_0 = 0$:

$$Ax_n = \sum_{i=1}^{n} \beta_{i-1} Ay_{i-1} = \sum_{i=1}^{n} (y_{i-1} - y_i) = y_0 - y_n = y - y_n ,$$

so that

$$r_n = y - Ax_n = y_n .$$

As a result, we now restate Theorem (II.2.2) in the following form.

<u>Theorem.</u> (21)

Let A be a bounded and positive bounded below operator on a

complex Hilbert space $\mathcal{H}$. If we define the sequence $\{x_n\}$ by

$$x_{n+1} = x_n + t_n r_n , \quad x_0 = 0 ,$$

where

$$t_n = \frac{\langle Ar_n, r_n\rangle}{\langle Ar_n, Ar_n\rangle} ,$$

then $x_n \to x^*$, the unique solution of $Ax=y$.

<u>Proof</u>.

We first observe that for $n \geq 0$:

$$r_{n+1} = y - Ax_{n+1} = y - A(x_n + t_n r_n) = r_n - t_n Ar_n$$

while $r_0 = y$. Thus

$$\|r_{n+1}\|^2 = \|r_n - t_n Ar_n\|^2 = \|r_n\|^2 - \frac{\langle Ar_n, r_n\rangle^2}{\|Ar_n\|^2} . \tag{22}$$

This means that the sequence $\{\|r_n\|\}$ is monotonically decreasing. Hence the sequence converges to some real number ρ , $0 \leq \rho \leq \|y\|$. This implies that

$$\lim_{n\to\infty} \frac{\langle Ar_n, r_n\rangle^2}{\|Ar_n\|^2} = 0 .$$

But now

$$\frac{\langle Ar_n, r_n\rangle^2}{\|Ar_n\|^2} \geq \frac{\gamma^4 \|r_n\|^4}{\|A\|^2 \|r_n\|^2} = \left(\frac{\gamma^2}{\|A\|}\right)^2 \|r_n\|^2 \geq 0 \tag{23}$$

for some $\gamma > 0$, so that $r_n \to 0$. Since A has a bounded inverse, $x_n \to x^*$. ▌

III.3

Because $\|x\| \le \frac{1}{\gamma^2} \|Ax\|$ for all $x \in H$, we have the following simple estimate for the error:

$$\|x_n - x^*\| \le \frac{1}{\gamma^2} \|Ax_n - y\| .$$

In [141] Sobolevskii gives the following definition.

<u>Definition.</u> (24)

Two linear operators P and R defined in a Hilbert space $\mathcal{H}$ <u>form an acute angle</u> if $D(P) = D(R)$; for some $\delta > 0$,

$$|\langle Pu, Ru \rangle| \ge \delta \|Pu\| \, \|Ru\| , \quad u \in D(P) ;$$

and $Pu=0$ and $Rv=0$ only when $u = v = 0$.

For specific properties of such operators the reader should consult Sobolevskii's article. Now we shall give the second method of Petryshyn [121]. We return to the original situation of two equations, equation (2) being easily solvable and equation (1) being the equation to be solved.

<u>Theorem.</u> (25)

Let the operator A be continuously invertible and form an acute angle with L . Assume that there exist real constants $\eta_3, \eta_4 > 0$ such that

$$\operatorname{Re}\langle Lu, Au \rangle \ge \eta_3 \|Au\|^2 , \quad \|Au\| \ge \eta_4 \|Lu\| .$$

If γ_2 is a real number such that $0 < \gamma_2 < 2$, if $x_0 \in D(A)$ is an arbitrary initial approximation to a solution of (1), and if

$Az_0 = Lx_0 - y'$, then the sequence $\{x_n\}$ determined by

$$x_{n+1} = x_n - \gamma_2 t_n z_n \ , \quad n=0,1,2,\ldots \ ,$$

where

$$t_n = \frac{\mathrm{Re}\langle Lz_n, Az_n\rangle}{\langle Lz_n, Lz_n\rangle} \ , \quad n=0,1,2,\ldots,$$

and

$$Az_{n+1} = Az_n - \gamma_2 t_n Lz_n \ , \quad n=0,1,2,\ldots,$$

has the property that $\|Lx_n - y'\| \to 0$ monotonically. If L is closed and $\|Lu\| \geq k\|u\|$ for every $u \in D(L)$ and some $k > 0$, then (1) has a unique solution x^* which is given by

$$x^* = x_0 - \gamma_2 \sum_{i=0}^{\infty} t_i z_i \ .$$

We do not give the proof here. It parallels the proof of Theorem (21). Notice that equation (17) remains unchanged so that $-r_n = Az_n$. Instead of (22) we obtain

$$\|Az_{n+1}\|^2 = \|Az_n\|^2 - \gamma_2(2-\gamma_2) \frac{(\mathrm{Re}\langle Lz_n, Az_n\rangle)^2}{\langle Lz_n, Lz_n\rangle} \ ,$$

which shows that the sequence $\{\|Az_n\|\}$, that is $\{\|r_n\|\}$, is monotonically decreasing and that

$$\lim_{n\to\infty} \frac{(\mathrm{Re}\langle Lz_n, Az_n\rangle)^2}{\langle Lz_n, Lz_n\rangle} = 0 \ .$$

The inequality corresponding to (23) is obtained from the hypotheses concerning η_3 and η_4 and from the acute angle properties, enabling one to show that $\|Az_n\| \to 0$.

Petryshyn gives the following error estimate for this method which follows immediately from the hypotheses of the second part of the theorem:

$$\|x_n - x^*\| \leq \frac{1}{k} \|Lx_n - y'\| .$$

Reasoning as before, one can see that $\gamma_2=1$ is the best choice here for the decrease of $\|Az_n\| = \| Lz_n - y' \|$. Using this choice, one can show that convergence is at least as fast as a geometric progression with ratio $q = (1-\delta^2)^{1/2} < 1$. (Notice that in Definition (24), $\delta > 0$ is necessarily less than or equal to one.)

In case L is a bounded and positive bounded below operator, we need only take $\gamma_2=1$, $x_0=0$, and $A=I$ to obtain Kolomý's Theorem (21) from Theorem (25).

In the article [121], Petryshyn gives applications of his two methods to both ordinary and partial differential equations.

SECTION 4

THE EXTENSION OF SOME ALGEBRAIC METHODS

Motivated by thoughts of advancing from solving n equations in n unknowns to solving an infinite number of equations in an infinite number of unknowns, it seems reasonable to attempt to carry certain methods of linear algebra over into the theory of operator equations. As we shall see in the last two chapters, this is true of the method of successive approximations and of the gradient methods. In this section we consider two methods. They are direct attempts to apply certain results of linear algebra to operator equations.

The first method we shall consider is due to Petryshyn [117, 119, 120, 126]. It generalizes the Gauss-Seidel and Jacobi methods of matrix theory [11] to a certain class of Kpd operators. Petryshyn calls this method the generalized overrelaxation iterative method or the go-method for short. The other method we wish to consider here is due to Kellogg [81]. It is a direct attempt at generalizing the Peaceman-Rachford method for matrix equations [35] to operator equations.

Let $\mathscr{H}_1$ and $\mathscr{H}_2$ be complex Hilbert spaces and let T be a bounded linear operator mapping $\mathscr{H}_1$ into $\mathscr{H}_2$. We define the following quantities:

$$m(T) = \inf_{\substack{u \in \mathcal{V}_1 \\ u \neq 0}} \frac{\|Tu\|}{\|u\|} , \quad M(T) = \sup_{\substack{u \in \mathcal{V}_1 \\ u \neq 0}} \frac{\|Tu\|}{\|u\|} .$$

These real numbers exist and are finite.

Lemma. (1)

Let $T : \mathcal{V} \to \mathcal{V}$ be a bounded linear operator. Then $m(T) > 0$ if and only if $T^{-1} : R(T) \to \mathcal{V}$ exists and is bounded. In this case $R(T)$ is closed in $\mathcal{V}$ and $m(T)M(T^{-1}) = 1$.

Proof.

The proof of the first assertion is well-known [41, p. 86]. Thus suppose that $m(T) > 0$. Let $y_n \in R(T)$, $y_n \to y$, $y_n = Tx_n$. Then $x_n = T^{-1}y_n$ and $\|x_n - x_m\| = \|T^{-1}y_n - T^{-1}y_m\| \leq \|T^{-1}\| \, \|y_n - y_m\| \to 0$. Thus $x_n \to x$ and $Tx_n \to Tx$. Hence $y_n \to Tx$ or $Tx = y$, and we conclude that $R(T)$ is closed in $\mathcal{V}$. Letting $Tx=y$ or $x=T^{-1}y$ we have

$$\frac{1}{m(T)} = \frac{1}{\inf\limits_{y \neq 0} \frac{\|y\|}{\|T^{-1}y\|}} = \sup_{y \neq 0} \frac{\|T^{-1}y\|}{\|y\|} = M(T^{-1})$$

so that $m(T)M(T^{-1}) = 1$. ▮

Now let $A : \mathcal{V} \to \mathcal{V}$ be a bounded linear operator and assume that for some bounded operator K there exists a constant $\alpha > 0$ such that

$$\langle Au, Ku \rangle \geq \alpha^2 \|Ku\|^2 , \quad u \in \mathcal{V} . \tag{2}$$

In this case we note that A need not be Kpd according to Definition (I.1.3) since we have not required K to have a bounded inverse on R(K) dense in $\mathcal{H}$.

We wish to present the go-method for solving the equation

$$Ax = y , \quad y \in \mathcal{H} , \tag{3}$$

where the operator A satisfies (2) for some K and is of the special form A=D-S-Q with the bounded operators D, S, and Q subject to the conditions:

(a) $\langle Du,Ku\rangle$ is real for all $u \in \mathcal{H}$.

(b) The operators $G_\omega = \frac{2-\omega}{\omega} D-S+Q^*$ satisfy an inequality similar to (2) for all $\omega > 0$.

(c) The equation $(D-\omega S)x=y$ is easily and uniquely solvable for every $y \in \mathcal{H}$, $\omega > 0$.

Lemma. (4)

For all $u, v \in \mathcal{H}$:

$$\langle Au,Kv\rangle = \langle Ku,Av\rangle , \quad \langle Du,Kv\rangle = \langle Ku,Dv\rangle .$$

Proof.

Inequality (2) shows that $\langle Ax,Kx\rangle$ is real for all $x \in \mathcal{H}$ while condition (a) gives the same for $\langle Dx,Kx\rangle$. Hence both results follow by using polarization identities, just as in the proof of Lemma (I.1.2).▌

One should note that this lemma will not hold in general in a real Hilbert space. If we assume the results of the lemma, then what follows is also valid in a real Hilbert space.

Lemma. (5)

Let N be a bounded and continuously invertible linear operator of the form $N=N_1-N_2$. Suppose that N_1 is bounded and continuously invertible and let $T=N_1^{-1}N_2$. If $\sigma(T) \subset \{z \in \mathbf{C} : |z| < 1\}$, then the sequence $\{v_n\}$ determined by

$$N_1 v_{n+1} = N_2 v_n + g \quad (\text{or } v_{n+1} = Tv_n + g_1), \tag{6}$$

with v_0 arbitrary and $g_1 = N_1^{-1}g$ converges to the solution v^* of $Nv=g$ for any $g \in \mathcal{H}$. The error is given by the estimate

$$\|v_{n+1}-v^*\| \le \|N^{-1}N_2\| \, \|v_n - v_{n+1}\|, \tag{7}$$

or more practically and less precisely by

$$\|v_{n+1}-v^*\| \le \frac{M(N_2)}{m(N)} \|v_n - v_{n+1}\|. \tag{8}$$

Proof.

The spectral radius, $r(T)$, is less than one. But $r(T) = \lim_{n\to\infty} \|T^n\|^{1/n}$ [41, p. 263], so by the root test $\sum_{n=0}^{\infty} \|T^n\|$ converges and $\|T^n\| \to 0$. To see that $\{v_n\}$ given by (6) converges to v^*, note that $D(N^{-1}) = R(N) = \mathcal{H}$ so that the solution v^* exists and is unique for each $g \in \mathcal{H}$. Let $e_n = v^* - v_n$. Then $e_n = Te_{n-1} = T^2e_{n-2} = \dots = T^ne_0$ so that $\lim_{n\to\infty} \|e_n\| \le \lim_{n\to\infty}\|T^n\| \, \|e_0\| = 0$.

Subtracting N_2v_{n+1} from both sides of (6):

$$Nv_{n+1} = N_2(v_n-v_{n+1}) + g = N_2(v_n-v_{n+1}) + Nv^*$$

or

$$(v_{n+1}-v^*) = N^{-1}N_2(v_n-v_{n+1})$$

from which we obtain (7). We obtain (8) from Lemma (1).▌

We shall use a fixed relaxation factor $\omega > 0$. Let $x_0 \in \mathcal{X}$ be arbitrary and define the sequence $\{x_n\}$ iteratively by

$$(D-\omega S)x_{n+1} = \{(1-\omega)D+\omega Q\}x_n + \omega y \ . \tag{9}$$

<u>Theorem.</u> (10)

Let A be a bounded operator of the form $A=D-S-Q$ satisfying (2), (a), (b), and (c). If K is positive bounded below and commutes with Q , and if $\sigma(T_\omega)$ contains only eigenvalues of finite multiplicity with zero as the only possible limit point where

$$T_\omega = (D-\omega S)^{-1}\{(1-\omega)D + \omega Q\} \ ,$$

then the sequence $\{x_n\}$ given by (9) converges to the unique solution x^* of (3). The error is given by

$$\|x_{n+1}-x^*\| \leq \|A^{-1}(\eta D-Q)\| \ \|x_{n+1}-x_n\| \ , \tag{11}$$

or more practically but less precisely by

$$\|x_{n+1}-x^*\| \leq \frac{M(\eta D-Q)}{m(A)} \|x_{n+1}-x_n\| \ , \tag{12}$$

where $\eta = 1 - \frac{1}{\omega}$.

Proof.

Since K is positive bounded below, K^{-1} exists and is bounded by Theorem (I.1.7). Thus, in this case, A is actually Kpd, and the theorem tells us that (3) has a unique solution x^*.

Write (9) in the form

$$x_{n+1} = T_\omega x_n + y_1$$

where $y_1 = \omega(D-\omega S)^{-1} y$. We shall use Lemma (5); hence we need only show that $\sigma(T_\omega) \subset \{z \in \mathbf{C} : |z| < 1\}$. By hypothesis, $\sigma(T_\omega)$ contains only eigenvalues $\lambda = \lambda(\omega)$ of T_ω, so it will be sufficient to show that $|\lambda| < 1$ where $T_\omega u = \lambda u$ for $u \neq 0$. We write this as

$$\{(1-\omega)D + \omega Q\}u = \lambda(D-\omega S)u \, . \tag{13}$$

Multiplying by -1 and adding $(D-\omega S)u$ to both sides of (13), we use $A=D-S-Q$ to obtain

$$\omega A u = (1-\lambda)(D-\omega S)u \, .$$

Since $\omega > 0$ and $u \neq 0$, the fact that A is Kpd shows that $\lambda \neq 1$. Thus $1-\lambda \neq 0$ and division yields

$$\frac{\omega}{1-\lambda} \langle Au, Ku \rangle = \langle (D-\omega S)u, Ku \rangle \tag{14}$$

by taking the scalar product with Ku.

Now take the scalar product of (13) with Ku to obtain

$$\omega \langle Qu, Ku \rangle = (\lambda + \omega - 1)\langle Du, Ku \rangle - \omega\lambda \langle Su, Ku \rangle \, .$$

Since D is K-symmetric (Lemma (4)), $\overline{\omega\langle Qu,Ku\rangle} = \omega\langle Ku,Qu\rangle = (\bar{\lambda}+\omega-1)\langle Du,Ku\rangle - \omega\bar{\lambda}\langle Ku,Su\rangle$, or $\omega\langle Q^*Ku,u\rangle = (\bar{\lambda}+\omega-1)\langle Du,Ku\rangle - \omega\bar{\lambda}\langle S^*Ku,u\rangle$. But A is also K-symmetric (Lemma (4)) and $S^* = D^*-A^*-Q^*$, so this may be written:

$$\omega\langle Q^*Ku,u\rangle + (1-\omega)\langle Du,Ku\rangle = \frac{\omega\bar{\lambda}}{1-\bar{\lambda}}\langle Au,Ku\rangle . \tag{15}$$

Adding (14) to (15):

$$\langle(2D-\omega D-\omega S)u,Ku\rangle + \omega\langle Q^*Ku,u\rangle = \omega\,\frac{1-|\lambda|^2}{|1-\lambda|^2}\langle Au,Ku\rangle$$

since

$$\frac{1}{1-\lambda} + \frac{\bar{\lambda}}{1-\bar{\lambda}} = \frac{1-|\lambda|^2}{|1-\lambda|^2} = f(\lambda) .$$

But K commutes with Q , hence also with Q^* , so we have

$$\langle G_\omega u,Ku\rangle = f(\lambda)\langle Au,Ku\rangle .$$

Inequality (2) and property (6) show that $f(\lambda) > 0$. But this means $|\lambda| < 1$, as required.

The error estimates (11) and (12) follow from (7) and (8), since (9) is equivalent to

$$N_1x_{n+1} = N_2x_n + \omega y$$

where $N_1=D-\omega S$ and $N_2=(1-\omega)D+\omega Q$ with $N=N_1-N_2=\omega A$. ▌

The presentation given here is due to Petryshyn [117]. In [119] Petryshyn generalizes Theorem (10) by showing that for A, D,

S, and Q satisfying conditions (a), (b), and (c), $\sigma(T_\omega) \subset \{z \in \mathbb{C} : |z| < 1\}$ if and only if A is Kpd. In view of Lemma (5), this means that the go-method converges to the unique solution x* of (3) if and only if A is Kpd. Further generalizations are given by Petryshyn in [126]. If we take $\omega=1$ in (9), we obtain a generalization of the classical Gauss-Seidel method, the choice K=I yielding the classical case. If we take $Q=S^*$, then we obtain a method which was first proven for operator equations by Krein and Prozorovskaja [96]. Their work was itself a generalization of a method due to Wessinger [145] for Fredholm operators. More recent work on this method, including error estimates, can be found in the article by Krasnoselskii and Stecenko [94]. Finally, the choice S=O (A=D-Q) yields a generalization of the classical Jacobi method. By taking K=I, one obtains the classical case. This is discussed by Petryshyn in [120], mainly for matrices, but the Hilbert space case is also mentioned.

A recent article by Koliha [83] is of interest in connection with the results of Petryshyn mentioned here. Although no really new methods for solving (3) are presented, Koliha does generalize and unify all of the methods mentioned above. The operator A is taken to be bounded and self-adjoint. (Koliha also discusses the case for Kpd operators, but for clarity we will use K=I here.) Let P and P^{-1} be bounded operators on $\mathscr{H}$. Then the equation

$$x = (I-PA)x + Py$$

is equivalent to (3). Koliha uses a successive approximation type iteration,

$$x_{n+1} = (I-PA)x_n + Py ,$$

to generate $\{x_n\}$. If A is positive bounded below, the method converges to the unique solution x^* of (3) or (16) for certain P. If A is merely positive bounded below on a closed subspace of $\mathcal{X}$, then the method converges if and only if (3) is solvable. Let $T=I-PA$. The method becomes well-defined only after a specific choice is made for the operator P.

In his main method Koliha has A in the form $A=D-S-Q$ with $S=L+L^*$ and D and Q self-adjoint. Let α and β be real constants such that $(\alpha D+\beta L)^{-1}$ exists. We now select P to be $(\alpha D+\beta L)^{-1}$. The operator P will be completely defined when α and β are chosen. In this situation T has the form $T = (\alpha D+\beta L)^{-1} \{(\alpha-1)D+(\beta+1)L+L^*+Q\}$. Now the choice $\alpha=1, \beta=0$ gives the Jacobi method; the choice $\alpha=1, \beta=-1$ gives the Gauss-Seidel method; and the choice $\alpha = \frac{1}{\omega}$, $\beta = -1$ gives the over-relaxation method. Koliha also replaces P by φP, $\varphi > 0$, to obtain an extrapolated form of these three methods. In effect, α and β are replaced by α/φ and β/φ respectively.

In Section IV.1 we shall give a theorem (Theorem IV.1.14)) which can be considered a generalization of the go-method. To see that this is so, one takes $C=D-\omega S$ and $B=\omega I$ in (IV.1.15). Then with A in the form $A=D-S-Q$, we have $L=(1-\omega)D+\omega Q$, and the iteration scheme (IV.1.15) coincides with (9).

Now we shall turn our attention to the method of Kellogg [81]. This technique was first given for matrices by Peaceman and Rachford [34] in 1955 and is now known as the Peaceman-Rachford method. Kellogg made the first attempt to generalize the method to operator equations. He calls the method the alternating direction method. A stronger formulation of Kellogg's result follows from a successive approximation theorem of de Figueiredo and Karlovitz [68]. We shall discuss this formulation in Section IV.4, after we have examined the theorem of de Figueiredo and Karlovitz. At this point we give the formulation by Kellogg.

<u>Definition.</u> (17)

A closed operator $B : \mathcal{H} \to \mathcal{H}$ with $D(B)$ dense in $\mathcal{H}$ is said to have <u>nonnegative real part</u> if both:

$$\mathrm{Re}\langle Bu,u\rangle \geq 0 , \quad u \in D(B) ,$$

and

$$\mathrm{Re}\langle B^*u,u\rangle \geq 0 , \quad u \in D(B^*) .$$

Since $B=B^{**}$ [41, p. 361 or 38, p. 305] we see that B has nonnegative real part if and only if B^* does. Hence the conclusions of the following lemma hold for B^* as well as for B.

<u>Lemma.</u> (18)

If $\rho > 0$ and B has nonnegative real part, then:

(a) $\rho I+B$ has a bounded inverse on $\mathcal{H}$. In fact $\|(\rho I+B)^{-1}\| \le 1/\rho$.

(b) The operator $T(B) = (\rho I-B)(\rho I+B)^{-1}$ has domain $\mathcal{H}$ and norm $\|T(B)\| \le 1$. The same is true for $T(B^*) = T(B)^*$.

(c) $I + T(B)$ is 1-1 .

(d) $R(I+T(B)) \subset D(B)$.

<u>Proof.</u>

(a) Given $u \in D(B)$ we have:

$$\rho\langle u,u\rangle = \langle \rho Iu,u\rangle \le \langle \rho Iu,u\rangle + \mathrm{Re}\langle Bu,u\rangle$$
$$= \mathrm{Re}\langle(\rho I+B)u,u\rangle \le \|(\rho I+B)u\|\ \|u\|$$

or

$$\rho\|u\| \le \|(\rho I+B)u\| .$$

Hence $(\rho I+B)^{-1}$ exists and is bounded on $R(\rho I+B)$ with the norm bounded by $1/\rho$. This means that $R(\rho I+B)$ is closed [41, p. 178]. Now note that because the same conclusions hold for the operator $(\rho I+B)^*$, we have $N((\rho I+B)^*) = 0$. But $N((\rho I+B)^*) = (R(\rho I+B))^{\perp}$ so $R(\rho I+B) = \mathcal{H}$.

(b) If $u \in \mathcal{H}$, then $v = (\rho I+B)^{-1}u$ belongs to $D(B)$ so $T(B)u = (\rho I-B)v$ is defined on all of $\mathcal{H}$. The inequality

$$\frac{\langle T(B)u,T(B)u\rangle}{\langle u,u\rangle} = \frac{\rho^2\langle v,v\rangle - 2\rho\mathrm{Re}\langle Bv,v\rangle + \langle Bv,Bv\rangle}{\rho^2\langle v,v\rangle + 2\rho\mathrm{Re}\langle Bv,v\rangle + \langle Bv,Bv\rangle} \le 1$$

shows that $\|T(B)\| \leq 1$. The same conclusion holds for $T(B^*)$, as we have commented. And:

$$\begin{aligned}
T(B^*) &= (\rho I-B^*)(\rho I+B^*)^{-1} \\
&= [2\rho I - (\rho I+B^*)](\rho I+B^*)^{-1} \\
&= 2\rho(\rho I+B^*)^{-1} - I \\
&= 2\rho((\rho I+B)^*)^{-1} - I \\
&= ((\rho I+B)^*)^{-1}[2\rho I-(\rho I+B)^*] \\
&= ((\rho I+B)^{-1})^*(\rho I-B)^* \\
&= [(\rho I-B)(\rho I+B)^{-1}]^* \\
&= T(B)^* .
\end{aligned}$$

(c) Suppose $(I+T(B))x=0$. Letting $v=(\rho I+B)^{-1}x$, we have $(\rho I+B)v + (\rho I-B)v = 0$ or $2\rho Iv = 0$. Therefore $v=0$ and $x=0$.

(d) Let $v \in R(I+T(B))$ so that $(I+T(B))u = v$ for some $u \in \mathcal{H}$. Then for $z = (\rho I+B)^{-1}u$, we have $(\rho I+B)z + (\rho I-B)z = v$ or $v=2\rho z$. But $z \in D(B)$, so $v \in D(B)$ also. ▌

Let A_1 and A_2 be unbounded operators in $\mathcal{H}$ with nonnegative real parts. Define an operator A by $A=A_1+A_2$ with domain $D(A) = D(A_1) \cap D(A_2)$. The operators A_1 and A_2 play the role which the operators H and V (for horizontal and vertical) play in the Peaceman-Rachford method. This motivates Kellogg's choice of name, alternating direction method. We shall solve the equation

$$Ax = y , \quad y \in \mathcal{H} , \tag{19}$$

iteratively. We define the alternating direction iterations:

$$u_n = (\rho I+A_1)^{-1}[(\rho I-A_2)v_n + y] , \tag{20}$$

$$v_{n+1} = (\rho I+A_2)^{-1}[(\rho I-A_1)u_n + y] . \tag{21}$$

For an arbitrary choice of $v_0 \in D(A_2)$, one obtains sequences $\{u_n\}$ with $u_n \in D(A_1)$ and $\{v_n\}$ with $v_n \in D(A_2)$. At this point we make the following assumptions:

(a) Equation (19) has a solution $x^* \in D(A)$.

(b) $A_1^* + A_2^*$ is 1-1 on its domain $D(A_1^*) \cap D(A_2^*)$.

We shall see that assumption (b) is not required by de Figueiredo and Karlovitz. Denote the operator $T(A_1)$ by T_1 and $T(A_2)$ by T_2 .

Lemma. (22)

If conditions (a) and (b) hold and $\rho > 0$, then both $R(I-T_1T_2)$ and $R(I-T_2T_1)$ are dense in $\mathcal{H}$.

Proof.

We prove only one part, the proof of the other part being similar. Assume $R(I-T_1T_2)$ is not dense in $\mathcal{H}$ and pick $u \neq 0$ such that $(I-T_1T_2)^*u=0$. This is possible because $(R(I-T_1T_2))^{\perp} =$ $= N((I-T_1T_2)^*)$. Let $v = T_1^*u$ so that $u = T_2^*v$. Now let $v_{-1} = (\rho I+A_2^*)^{-1}v$ and $u_{-1} = (\rho I+A_1^*)^{-1}u$. Then by (a) of Lemma (18), $u_{-1} \in D(A_1^*)$ and $v_{-1} \in D(A_2^*)$ and

$$u = \rho u_{-1} + A_1^* u_{-1} , \tag{23}$$

$$v = \rho v_{-1} + A_2^* v_{-1} , \tag{24}$$

$$u = T_2^* v = \rho v_{-1} - A_2^* v_{-1} , \tag{25}$$

$$v = T_1^* u = \rho u_{-1} - A_1^* u_{-1} . \tag{26}$$

Adding (23) to (26) and (24) to (25), we obtain:

$$u + v = 2\rho u_{-1} ,$$

$$u + v = 2\rho v_{-1} .$$

Hence $u_{-1} = v_{-1} \in D(A_1^*) \cap D(A_2^*)$. Subtracting (25) from (23):

$$A_1^* u_{-1} + A_2^* u_{-1} = 0 .$$

Property (b) implies that $u_{-1}=0$. But this means that $u=0$, a contradiction. ▌

We shall pause here to prove a lemma due to de Figueiredo and Karlovitz [68] while the current situation is still fresh in our minds. This technical lemma will be of use when we discuss the results of de Figueiredo and Karlovitz in Section IV.4. In this lemma we do not require that assumptions (a) and (b) hold. Define an operator T by $T=T_1T_2$.

Lemma. (27)

Let $\mathcal{H}$, A_1, A_2, T_1, T_2, T, and $\rho > 0$ be as defined above. Then $x^* \in D(A)$ and x^* solves (19) if and only if $u^* = (\rho I + A_2)x^*$ solves

$$x - Tx = (I+T_1)y \ , \quad y \in \mathcal{H} \ . \tag{28}$$

Proof.

Assume that $x^* \in D(A)$ is such that x^* solves (19). Then:

$$\begin{aligned} u^* - Tu^* &= (\rho I+A_2)x^* - T(\rho I+A_2)x^* \\ &= (\rho I-A_1)x^* + A_1x^* + A_2x^* - T_1(\rho I-A_2)x^* \\ &= T_1(\rho I+A_1)x^* + Ax^* - T_1(\rho I-A_2)x^* \\ &= T_1A_1x^* + Ax^* + T_1A_2x^* \\ &= (I+T_1)Ax^* \\ &= (I+T_1)y \ . \end{aligned} \tag{29}$$

Conversely, assume that $u^*=(\rho I+A_2)x^*$ solves (28). Then $x^* \in D(A_2)$, and we shall show that $x^* \in D(A_1)$, implying that $x^* \in D(A)$. First:

$$\begin{aligned} u^* - Tu^* &= u^* - (\rho I-A_1)(\rho I+A_1)^{-1}(\rho I-A_2)(\rho I+A_2)^{-1} u^* \\ &= (I+T_1)y \end{aligned}$$

or

$$u^* - (\rho I-A_1)w' = (I+T_1)y \ ,$$

where

$$w = (\rho I-A_2)x^* \quad \text{and} \quad w' = (\rho I+A_1)^{-1}w \ .$$

Thus we have both

$$u^* + w = (\rho I + A_2)x^* + (\rho I-A_2)x^* = 2\rho x^*$$

and

$$u^* + w = [(I+T_1)y + (\rho I-A_1)w'] + (\rho I+A_1)w'$$
$$= 2\rho w' + (I+T_1)y \ .$$

These combine to produce

$$x^* = w' + \frac{1}{2\rho}(I+T_1)y \ .$$

We note that $w' \in D(A_1)$ by definition. By part (d) of Lemma (18), $(I+T_1)y \in R(I+T_1) \subset D(A_1)$. Thus $x^* \in D(A_1)$.

Finally, as in (29):

$$(I+T_1)y = u^* - Tu^* = (I+T_1)Ax^* \ .$$

But part (c) of Lemma (18) shows that $I+T_1$ is 1-1 . Thus $Ax^*=y$ or x^* solves (19).▌

Now we define the averages:

$$U_n = \frac{1}{n+1}(u_0 + u_1 + u_2 + \ldots + u_n) \ ,$$

$$V_n = \frac{1}{n+1}(v_0 + v_1 + v_2 + \ldots + v_n) \ .$$

Theorem. (30)

If conditions (a) and (b) hold and $\rho > 0$, then for any choice of $v_0 \in D(A_2)$ the sequences $\{U_n\}$ and $\{V_n\}$ both converge to x^* .

Proof.

We shall show that $V_n \to x^*$; the proof of the other assertion is similar. Since $u_n \in D(A_1)$ and $v_n \in D(A_2)$, we may define $r_n = (\rho I+A_1)u_n$ and $s_n = (\rho I+A_2)v_n$. Using (20) and (21) we obtain:

$$r_n = T_2 s_n + y$$

$$s_{n+1} = T_1 r_n + y . \tag{31}$$

Thus $s_{n+1} = T_1 T_2 s_n + (I+T_1)y$. Similarly we define $r=(\rho I+A_1)x^*$ and $s=(\rho I+A_2)x^*$. Noting that the formulas in (31) hold with r in place of r_n and s in place of s_n and s_{n+1}, we obtain $s=T_1T_2 s + (I+T_1)y$. Thus

$$s - s_{n+1} = T(s-s_n) , \tag{32}$$

where $T=T_1T_2$. If we define $S_n=(\rho I+A_2)V_n$, then

$$S_n = \frac{1}{n+1}(s_0 + s_1 + \dots + s_n) ,$$

and we obtain using (32):

$$\begin{aligned} s - S_n &= \frac{1}{n+1}[(s-s_0) + (s-s_1) + \dots + (s-s_n)] \\ &= \frac{1}{n+1}[(s-s_0) + T(s-s_0) + \dots + T^n(s-s_0)] \\ &= \frac{1}{n+1}[I + T + T^2 + \dots + T^n](s-s_0) . \end{aligned} \tag{33}$$

Let $\epsilon > 0$ be arbitrary. By Lemma (22) we can find $w \in \mathcal{X}$ such that

$$\|(s-s_0) - (I-T)w\| < \frac{\epsilon}{2} .$$

From part (b) of Lemma (18): $\|T\| \le \|T_1\| \, \|T_2\| \le 1$. Writing $s-s_0$ in the form $(I-T)w + [(s-s_0) - (I-T)w]$ in (33):

$$\begin{aligned}\|s-S_n\| &\le \frac{1}{n+1} \|(I+T+T^2 + \ldots + T^n)(I-T)w\| + \left(\frac{n+1}{n+1}\right)\frac{\epsilon}{2} \\ &= \frac{1}{n+1} \|(I-T^{n+1})w\| + \frac{\epsilon}{2} \\ &\le \frac{2}{n+1} \|w\| + \frac{\epsilon}{2} .\end{aligned}$$

Therefore, for n sufficiently large, $\|s-S_n\| < \epsilon$. Now using part (a) of Lemma (18):

$$\begin{aligned}\|x^*-V_n\| &= \|(\rho I+A_2)^{-1}(s-S_n)\| \\ &\le \frac{1}{\rho} \|s-S_n\| .\end{aligned}$$

Hence $V_n \to x^*$. ▮

An even better result can be obtained if, instead of assumptions (a) and (b), we make the stronger assumption that A_1 and A_2 are self-adjoint and positive bounded below. In this case we shall see that we have convergence of the two sequences $\{u_n\}$ and $\{v_n\}$ directly to the unique solution x^* of (19), not just the convergence of their averages. Observe that it would be possible merely to require that A_1 and A_2 be positive bounded below, if we used Theorem (I.1.7) to extend them to self-adjoint operators.

<u>Theorem.</u> (34)

Let the operators A_1 and A_2 be self-adjoint and positive bounded below. If $0 < \rho < \alpha$, where $\alpha = \inf\{\sigma(A_1) \cup \sigma(A_2)\}$, then for any choice of $v_0 \in D(A_2)$ the sequences $\{u_n\}$ and $\{v_n\}$ given by (20) and (21) both converge to the unique solution x^* of (19).

Proof.

We have $\sigma(A_i) \subset [\alpha,\infty)$ for $i=1,2$. The functions $f(\lambda) = \frac{\lambda-\rho}{\lambda+\rho}$ and $g(\lambda) = \frac{1}{f(\lambda)}$ are bounded and positive on $[\alpha,\infty)$ and $0 < f(\lambda) < 1$. From the spectral theorem we are able to conclude that the operators $-T_i=f(A_i)$ and $(-T_i)^{-1}=g(A_i)$ are bounded, self-adjoint, and positive definite as are their square roots for $i=1,2$. Note that $\|(-T_i)^{1/2}\| < 1$ for $i=1,2$. Now define a new inner product on $D(T_2) = \mathcal{H}$ by

$$\langle v,w\rangle_2 = -\langle T_2 v,w\rangle ,$$

with the associated norm denoted by $\|\cdot\|_2$. We claim that these norms are equivalent:

$$\|(-T_2)^{-1/2}\|^{-1}\|v\| = \|(-T_2)^{-1/2}\|^{-1}\langle(-T_2)^{-1/2}(-T_2)^{1/2}v,(-T_2)^{-1/2}(-T_2)^{1/2}v\rangle^{1/2}$$

$$\leq - \langle T_2 v,v\rangle^{1/2} = \|v\|_2 ,$$

and

$$\|v\|_2 = - \langle T_2 v,v\rangle^{1/2} = \langle(-T_2)^{1/2}v, (-T_2)^{1/2}v\rangle^{1/2} = \|(-T_2)^{1/2}v\|$$

$$\leq \|(-T_2)^{1/2}\| \, \|v\| .$$

Using the norm $\|\cdot\|_2$ we obtain a new Hilbert space $\mathcal{H}_2$. For

$T = T_1T_2$ we have

$$\langle Tv,w\rangle_2 = -\langle T_2T_1T_2v,w\rangle = -\langle T_2v,Tw\rangle = \langle v,Tw\rangle_2 ,$$

so that T is self-adjoint on $\mathcal{H}_2$. If $v \neq 0$, then:

$$\begin{aligned}\|Tv\|_2^2 &= \langle(-T_2)T_1(-T_2)v, T_1(-T_2)v\rangle \\ &= \|(-T_2)^{1/2}T_1(-T_2)^{1/2}(-T_2)^{1/2}v\|^2 \\ &< \|(-T_2)^{1/2}v\|^2 = \|v\|_2^2 .\end{aligned}$$

This shows that in the space $\mathcal{H}_2$, $\|T\|_2 \leq 1$ and -1 is not an eigenvalue of T. Thus we can use a theorem of Krasnoselskii [93], which we shall prove in Chapter IV (Theorem (IV.2.1)), to show that

$$\lim_{n\to\infty} \|T^n v\|_2 = 0 . \tag{35}$$

Krasnoselskii shows that, under these conditions, the sequence $\{x_n\}$ given by $x_{n+1} = Tx_n + v$ converges. Since $x_n = T^n x_0 + \sum_{i=0}^{n-1} T^i v$, the series $\sum_{i=0}^{\infty} T^i v$ must converge. This implies (35).

Now, using (32), we have

$$\|s-s_{n+1}\|_2 = \|T(s-s_n)\|_2 = \|T^2(s-s_{n-1})\|_2 = \ldots = \|T^{n-1}(s-s_0)\|_2 \to 0$$

from (35). The equivalence of norms insures that $s_n \to s$ in $\mathcal{H}$. That is, $(\rho I+A_2)v_n \to (\rho I+A_2)x^*$. The operator $\rho I+A_2$ is continuously invertible by part (a) of Lemma (18). Hence $v_n \to x^*$. In a similar manner one shows that $u_n \to x^*$. ∎

III.4

CHAPTER IV

SUCCESSIVE APPROXIMATION METHODS

SECTION 1

INTRODUCTION

Loosely speaking, any iterative method is a successive approximation method. That is, we use these methods to obtain iteratively a sequence of approximations $\{x_n\}$ to the solution of the equation

$$Ax = y \ , \quad y \in \mathcal{Y} \ . \tag{1}$$

However, we have in mind here the classical method of successive approximations, associated with the Neumann series, traditionally used to solve an equation of the second kind. Therefore, we shall take $A=I-T$ and consider equation (1) in the form

$$x - Tx = y \ , \quad y \in \mathcal{Y} \ . \tag{2}$$

The basic recursion formula is given by

$$x_{n+1} = Tx_n + y \ , \quad x_0 \ \text{arbitrary.} \tag{3}$$

Notice that this may be written in the form

$$x_{n+1} = T^{n+1}x_0 + \sum_{i=0}^{n} T^i y \ . \tag{4}$$

The term at the right is a partial sum of the infinite series

$$\sum_{i=0}^{\infty} T^i y \, , \tag{5}$$

which is called the Neumann series [33]. Observe that if the sequence given by (3) or (4) converges, then it must converge to a solution of (2). We shall show in Section 2 that, under certain conditions, even if only a subsequence $\{x_{n_j}\}$ of $\{x_n\}$ converges, the sequence itself must converge to a solution of (2). In this chapter we shall consider the method of successive approximations as explained above and recent improvements associated with it. It will often be natural to consider these methods in a Banach space. An early presentation of this method in the context of functional analysis, both for linear and nonlinear equations, may be found in the 1939 work of Kantorovic [25].

As we stated in Chapter I, we shall not pursue the general subject of fixed point theorems. However, one should note that the fixed points of the function $\hat{T}$ given by $\hat{T}(x) = Tx+y$ coincide with the solutions of (2). Thus, in certain cases, we shall find it expedient to consider certain fixed point theorems in order to arrive at the solutions of (2).

In Section 1 of this chapter, we shall examine the basic method of successive approximations. A few historical comments will be given, and we shall refer to certain results which will not be discussed in detail. In Section 2 we consider an important Banach space result of Browder and Petryshyn [56] which generalizes an older Hilbert space result of Krasnoselskii [93]. Also we shall consider a Hilbert space method of Petryshyn [123], based on the result of Browder and Petryshyn, which replaces 2^n steps of the ordinary recursion formula (3) by only n steps, thus accelerating convergence. In Section 3 we shall discuss a fixed point theorem of Krasnoselskii [92] and some of its generalizations due to Schaefer [137], Reinermann [131], and Oblomskaja [113]. Finally, in Section 4 we discuss a Banach space result of de Figueiredo and Karlovitz [68] which generalizes the result of Browder and Petryshyn discussed in Section 2. As a corollary to this result, we obtain an improved formulation of the Hilbert space result of Kellogg given in Section III.4.

At this point we wish to prove a very basic theorem for the method of successive approximations.

Theorem. (6)

Let T be a linear operator on a Banach space $\mathcal{B}$ such that $\|T\| < 1$. Then the equation (2) has a unique solution x^* for every $y \in \mathcal{B}$ given by (5). That is, the sequence $\{x_n\}$ given by (3) or (4) converges to x^*.

Proof.

Since $\|T\| < 1$ and $\|T^n\| \leq \|T\|^n$, the series $\sum_{i=0}^{\infty} T^i$ converges to some bounded operator S. But

$$ST = TS = \sum_{i=0}^{\infty} T^{i+1} = \sum_{i=0}^{\infty} T^i - I = S - I ,$$

so that

$$(I-T)S = S(I-T) = I .$$

Thus $S=(I-T)^{-1}$ and $x^*=(I-T)^{-1}y = \sum_{i=0}^{\infty} T^i y$. Since $\|T^{n+1}x_0\| \leq \|T\|^{n+1}\|x_0\| \to 0$, formula (4) shows that $x_n \to x^*$. ▌

This method is commonly called the method of Picard-Poincaré-Neumann. Historical justification for this name is given by Hellinger and Toeplitz [20, pp. 1346-1354]. They point out that the earliest use of the method in conjunction with integral equations is due to Liouville [28]. The later application to integral equations by Neumann [33] is reproduced in Courant and Hilbert [6, vol. I, pp. 140-142]. Of course, the method hardly originated with Liouville. In fact, in his discussion of the method, Liouville states [28, p. 21], "... continuant indéfiniment cette opération, conformément à la méthode connue des approximations successives, on a enfin ...". This clearly indicates that the method was well-known in some context at the time of Liouville. Rall [129] points out that Heron of Alexandria used the method of successive approximations in the second century B.C. to solve $x^2=3$. His recursion formula, based on the fact that $x = \frac{1}{2}(x + \frac{3}{x})$ reduces to $x^2=3$, was of the form $x_{n+1}= \frac{1}{2}(x_n+ \frac{3}{x_n})$. Thus the actual origins of the method of successive approximations are no doubt lost in antiquity.

We now observe that Theorem (6) shows that

$$\sum_{i=0}^{\infty} T^i = (I-T)^{-1} = A^{-1}$$

when $\|T\| < 1$. This bears a close resemblance to, and was no doubt motivated by, the geometric series $\sum_{i=0}^{\infty} a^i = (1-a)^{-1}$ for $|a| < 1$. We now look at a slightly more general use of the Neumann series, which is again motivated by the geometric series. For the operator $I-T=A$ to have an inverse in the Banach space $\mathcal{B}$, we must at least be able to find some approximate inverse S . This operator S must be "close enough" to A^{-1} in the sense that SA is "close" to the identity operator I . We make this precise by requiring that $\|I-SA\|$ be small, say less than one. If $\|T\| < 1$ we may select $S=I$, for then $\|I-SA\| = \|I-S(I-T)\| = \|T\| < 1$. This is the case considered in Theorem (6). The following is due to Rall [128, 129].

<u>Theorem.</u> (7)

Let $T : \mathcal{B} \to \mathcal{B}$ be a bounded linear operator. Then $(I-T)^{-1} = A^{-1}$ exists if and only if there exists a bounded linear operator S on $\mathcal{B}$ such that S^{-1} exists and

$$\|I-SA\| < 1 . \tag{8}$$

In this case

$$A^{-1} = \sum_{i=0}^{\infty} (I-SA)^i S . \tag{9}$$

We shall not prove this theorem since it is a special case of a result of Petryshyn which we are about to consider. Notice that if such an approximate inverse can be found, then in place of (3), we may use

$$x_{n+1} = (I-SA)x_n + Sy$$

to generate a sequence $\{x_n\}$ of successive approximations. Notice also that if, instead of the identity $(1-a)^{-1} = \sum_{i=0}^{\infty} a^i$, we consider the identity $(1-a)^{-1} = \left(\sum_{i=0}^{n-1} a^i\right) \cdot \left(\sum_{i=0}^{\infty} a^{in}\right)$, where $|a^n| < 1$, then we see that condition (8) can be replaced by the condition $\|(I-SA)^n\| < 1$ for some natural number n. In particular, this condition always holds when the operator $I-SA$ is quasinilpotent.

We see from Theorem (6) that the Neumann series (5) converges whenever $\|T\| < 1$. Using the fact that $r(T)$, the spectral radius of T, is equal to $\lim_{n\to\infty} \|T^n\|^{1/n}$, the root test shows that (5) converges whenever $r(T) < 1$. For the most part, the rest of this chapter is devoted to looking at conditions weaker than $\|T\| < 1$ or $r(T) < 1$ which will allow us to conclude that the method of successive approximations converges. For the moment we wish to look at a near converse to the statement that $r(T) < 1$ implies that the series (5) converges. This is due to Reinermann [131].

<u>Theorem.</u> (10)

Let T be a bounded linear operator on a Banach space $\mathcal{B}$.

If the Neumann series (5) converges for every $y \in \mathcal{B}$, then $r(T) \leq 1$.

Proof.

Define linear operators B_n by

$$B_n y = \sum_{i=0}^{n} T^i y .$$

Then $\|B_n y\| \leq N$ by hypothesis, where N depends on y. By the Uniform Boundedness Principle, we can find M such that $\|B_n\| \leq M$ for all n. For $n \geq 1$ we have

$$\|T^n\| = \|\sum_{i=0}^{n} T^i - \sum_{i=0}^{n-1} T^i\| \leq \|\sum_{i=0}^{n} T^i\| + \|\sum_{i=0}^{n-1} T^i\|$$

$$= \|B_n\| + \|B_{n-1}\| \leq 2M .$$

Thus

$$r(T) = \lim_{n\to\infty} \|T^n\|^{1/n} \leq \lim_{n\to\infty} (2M)^{1/n} = 1 . \blacksquare$$

One might hope that a true converse were available for the statement that $r(T) < 1$ implies the series (5) converges. That is, we would wish to change the conclusion of Theorem (10) to $r(T) < 1$. However, this is not possible as we shall see in Example (2.4).

If T is completely continuous, then we can strengthen the conclusion of Theorem (10) to $r(T) < 1$. For suppose that $\lambda \in \sigma(T)$, $\lambda \neq 0$. Then $Tx = \lambda x$ for some $x \neq 0$ and we may assume that $\|x\| = 1$. By induction $T^n x = \lambda^n x$ for $n=1,2,\dots$. Now $\sum_{n=0}^{\infty} T^n x$ converges so $\|T^n x\| \to 0$. Thus $\lim_{n\to\infty} \|\lambda^n x\| = \lim_{n\to\infty} |\lambda^n| = \lim_{n\to\infty} |\lambda|^n = 0$ or $|\lambda| < 1$. But $\sigma(T)$ is compact, so $r(T) < 1$.

To conclude this section, we shall prove a Banach space result of Petryshyn [118] which, under certain conditions, generalizes the three methods of Bialy given in Section III.2. The theorem is concerned with solving (1) in a Banach space $\mathcal{B}$ when A is a bounded linear operator. The technique used is a variation of the concept of visualizing (1) in the form $(I-T)x = [I-(I-A)]x = y$ and applying the method of successive approximations. In order to prove this theorem, we must first prove a substantial preliminary result from which the main result will follow rather quickly. After the proof of this preliminary theorem, we shall state and prove a corollary. Then we shall prove the main result.

Theorem. (11)

Equation (1) has a unique solution x^* for every $y \in \mathcal{B}$ if and only if there exist a continuously invertible operator C and a bounded linear operator B with an inverse B^{-1} having the property that the series

$$\sum_{i=0}^{\infty} T^i g \tag{12}$$

converges for every $g \in \mathcal{B}$ where $T=C^{-1}L$ and $L=C-BA$. In this case x^* is given by (12) with $g=C^{-1}By$.

Proof.

Suppose that (1) has a unique solution x^* for every $y \in \mathcal{B}$. This means that A is continuously invertible. Let C be an arbitrary continuously invertible operator on $\mathcal{B}$ and put $B=\eta CA^{-1}$ where $\eta \in \mathbb{C}$ is such that $|1-\eta| < 1$. This choice insures that $B^{-1} = \frac{1}{\eta} AC^{-1}$ exists and $T=C^{-1}L=(1-\eta)I$ has $\|T\| < 1$ Theorem (6) shows that (12) converges for all $g \in \mathcal{B}$. In particular, it converges with $g=C^{-1}By$ to the unique solution x^* of $(I-T)x = C^{-1}By$ which is equivalent to (1).

Conversely, let C and B be as stated in the theorem and assume that (12) converges for every $g \in \mathcal{B}$. Then it converges for $g=C^{-1}By$ for every $y \in \mathcal{B}$. Note that it converges to some x^* which solves (1), for $Tx^*=x^*-g$ implies that $C^{-1}BAx^*=C^{-1}By$ which is equivalent to (1). We need only show that x^* is unique. Suppose $(I-T)v=0$. Then $v=Tv=T^2v=\ldots$ and passing to the limit in

$$v = \frac{1}{n+1} \sum_{i=0}^{n} T^i v$$

shows that $v=0$ or $(I-T)^{-1}$ exists. If $\bar{x}$ also solves (1), then for $v=\bar{x}-x^*$ we have $C^{-1}BAv=0$ or $(I-T)v=0$. Thus $v=0$ or $\bar{x}=x^*$. ▌

Petryshyn points out that the verification of condition (12) may be difficult in actual practice. Thus he proves a corollary which gives successively easier (but also successively more restrictive) conditions that can be used in place of condition (12).

Corollary. (13)

The assertions of Theorem (11) remain valid if condition (12) is replaced by any one of:

(a) $\limsup_{n\to\infty} \|T^n g\|^{1/n} < 1$ for every $g \in \mathcal{B}$.

(b) $\lim_{n\to\infty} \|T^n\|^{1/n} < 1$.

(c) $\|T\| < 1$.

Proof.

We know that $r(T) = \lim_{n\to\infty} \|T^n\|^{1/n}$ and $r(T) \leq \|T\|$. Thus condition (c) implies condition (b). Because $\|T^n g\|^{1/n} \leq \|T^n\|^{1/n}\|g\|^{1/n}$ and because $\|g\|^{1/n} \to 1$ we see that condition (b) implies condition (a). Finally, if (a) holds, then $\sum_{i=0}^{\infty} \|T^i g\|$ converges by the root test. Thus

$$\|\sum_{i=0}^{n+p} T^i g - \sum_{i=0}^{n} T^i g\| = \|\sum_{i=n+1}^{n+p} T^i g\| \leq \sum_{i=n+1}^{n+p} \|T^i g\| < \epsilon$$

for n sufficiently large and all p. Hence (12) converges for all $g \in \mathcal{B}$. ∎

Now we can state and prove the theorem of Petryshyn in which we are interested.

Theorem. (14)

If (1) has a unique solution $x^* \in \mathcal{B}$ for every $y \in \mathcal{B}$, then the sequence $\{x_n\}$ given by

$$Cx_{n+1} = Lx_n + By , \qquad (15)$$

where C is continuously invertible, B is bounded and 1-1 , and $L=C-BA$, converges to x^* for any choice of $x_0 \in \mathcal{B}$ if and only if

$$\sum_{i=0}^{\infty} (C^{-1}L)^i x$$

converges for every $x \in \mathcal{B}$.

Proof.

Let $\{x_n\}$ be given by (15) with $x_0 \in \mathcal{B}$ arbitrary. Equation (15) may be written

$$x_{n+1} = Tx_n + g ,$$

where $g=C^{-1}By$ and $T=C^{-1}L$. Using a representation similar to (4) we see that if (12) converges for all $g \in \mathcal{B}$, then taking $g=x_0$, $T^{n+1}x_0 \to 0$ and $\{x_n\}$ must converge for all $g \in \mathcal{B}$. On the other hand, if for any $g \in \mathcal{B}$, $\{x_n\}$ converges for any choice of x_0 , then it converges for $x_0=0$. Thus $x_n = \sum_{i=0}^{n-1} T^i g$ shows that (12) converges for all $g \in \mathcal{B}$. We now apply Theorem (11) and note that $x^* = \sum_{i=0}^{\infty} T^i g$ with $g=C^{-1}By$ independent of the choice of x_0 , because $\lim_{n\to\infty} T^n x_0=0$ for any $x_0 \in \mathcal{B}$. ▌

A corollary similar to Corollary (13) can be given here. For the reader interested in Banach space results, it is of interest to note that Theorem (14) generalizes not only Bialy's and Rall's results, but also those of many others including Fridman, Bückner, Cesari and Picone, von Mises and Polaczek-Geringer, Keller, Quade, Wiarda, Wagner, and Samuelson. In addition it generalizes the Hilbert space method of Petryshyn, called the go-method, which we discussed in Section III.4.

In [118] Petryshyn gives error estimates for his method, which we do not discuss. Improved error estimates are given in [122]. A recent article by Mosolov [109] contains a slight generalization of Theorem (11).

Now assume that $\mathcal{B}$ is a Hilbert space. Let $C=I$ and $B=\alpha I$, $\alpha > 0$, in Theorem (14). Suppose that A is positive. The analog of condition (c) above shows that it suffices to require that $\|I-\alpha A\| < 1$ which, because A is positive, implies that $0 < \alpha < \frac{2}{\|A\|}$ [38, p. 265]. In this case the iterative scheme of Theorem (14) reduces to that used in Theorem (III.2.6) of Bialy. Note that (1) is uniquely solvable by Theorem (6). It was pointed out that Bialy's recursion formula (III.2.7) could be written in the form

$$x_{n+1} = (I-\alpha A)x_n + \alpha y \ .$$

If we apply Theorem (III.2.6) to the equation $A^*Ax=A^*y$, which is equivalent to (1), assuming (1) is solvable (Theorem (I.1.8)), this recursion formula takes on the form

$$x_{n+1} = (I-\alpha A^*A)x_n + \alpha A^*y .$$

This is equivalent to taking $C = \frac{1}{\alpha} I$ and $B=A^*$ in Theorem (14) with $0 < \alpha < \frac{2}{\|A^*A\|} = \frac{2}{\|A\|^2}$. In this case the iterative scheme of Theorem (14) is reduced to that used in Theorem (III.2.11), provided (1) is uniquely solvable.

Finally, let $C = \frac{1}{\alpha^2} I$ and $B=A$ in Theorem (14), and require A to be self-adjoint. Then (15) takes the form

$$x_{n+1} = x_n + \alpha^2 Ar_n . \tag{16}$$

We have shown in the proof of Theorem (III.2.15) that $x_{n+1} = x_{n-1} + \alpha^2 Ar_{n-1}$. Thus the subsequences $\{x_{2n}\}$ and $\{x_{2n-1}\}$ of Bialy's Theorem (III.2.15) coincide with some sequence given by recursion formula (16) and show the relationship between Theorem (III.2.15) and Theorem (14).

A basic method for solving (2) is due to Sokolov [142]. It is sometimes called Sokolov's method, but, especially in the Russian literature, it is usually referred to as the method of averaging functional corrections. A detailed presentation of the method is given in the book by Lucka [104]. We do not discuss this method. However, we do wish to mention a variation of the method by Peklova [115] since it combines the techniques of both the method of successive approximations and the method of steepest descent. Peklova calls the method the gradient variant of Sokolov's method. Instead of using (3) to generate the sequence $\{x_n\}$, Peklova uses

$$x_{n+1} = T(x_n + \alpha_{n+1} v_n) + y \ .$$

Here T is a bounded linear operator on a real Hilbert space such that I-T=A is continuously invertible, and

$$\alpha_{n+1} v_n = \frac{\langle x_{n+1} - x_n, v_n \rangle}{\langle v_n, v_n \rangle} v_n \ .$$

The sequence $\{v_n\}$ may be chosen in several ways, but Peklova's main interest is in the choice $v_n = r_n = y-(I-T)x_n = y-Ax_n$. The connection between this variation and the method of steepest descent will become clear in Chapter V.

Finally, we mention two articles by Kurpel. In [97] Kurpel deals with solving the equation (2) in a Hilbert space. Instead of (3), he uses

$$x_{n+1} = (I-P)Tx_n + PT(I-T)y \ ,$$

where P is a certain projection operator. Sufficient conditions for convergence are given. In [98] Kurpel extends his results to Banach spaces. We do not discuss Kurpel's technique in this paper.

SECTION 2

A THEOREM OF BROWDER AND PETRYSHYN

In this section we shall state and prove an important Banach space theorem due to Browder and Petryshyn [56]. This theorem is a generalization of both Theorem (1.6) and of a Hilbert space result due to Krasnoselskii [93]. We conclude the section with a Hilbert space method of Petryshyn which replaces 2^n steps of the ordinary method of successive approximation by only n steps. This method is actually a corollary to the result of Browder and Petryshyn. Also, the method itself has a corollary which gives a very general Hilbert space technique for solving the equation of the first kind for any bounded operator.

As we mentioned in Section 1, we are trying to find ways in which the requirement $\|T\| < 1$ of Theorem (1.6) can be weakened. One solution to this problem was given in 1960 by Krasnoselskii [93]. He is able to use the hypothesis $\|T\| \leq 1$, but other restrictions on T are added. We state his theorem.

Theorem. (1)

Let T be a bounded and self-adjoint operator on a complex Hilbert space $\mathcal{H}$ such that $\|T\| \leq 1$ and such that -1 is not an eigenvalue of T. If the equation

$$(I-T)x = y , \quad y \in \mathcal{H} , \tag{2}$$

has a solution (not necessarily unique) for a certain $y \in \mathcal{H}$,
then the successive approximation sequence $\{x_n\}$ given by

$$x_{n+1} = Tx_n + y, \quad x_0 \text{ arbitrary,} \tag{3}$$

converges to a solution of equation (2).

We shall not prove this theorem now as it is a special case of the theorem we are about to prove. Krasnoselskii shows that if P is the projection of $\mathcal{H}$ onto $N(I-T)$, then the sequence $\{x_n\}$ given by (3) converges to x^*+Px_0, where x^* is a solution of equation (2), orthogonal to $N(I-T)$. We comment that Krasnoselskii's proof uses the spectral theorem while the result of Browder and Petryshyn has a more elementary proof.

At this point we wish to give an example of an operator to which Theorem (1) can be applied, but to which Theorem (1.6) cannot be applied. This example will demonstrate the applicability of Krasnoselskii's result, and it will also provide an example for the claim made in Section 1 that the conclusion of Theorem (1.10) can not be strengthened to read $r(T) < 1$.

Example. (4)

Let $L^2[0,1]$ be the Hilbert space of equivalence classes of square integrable, complex-valued functions on $[0,1]$. Define an operator T by $Tx(t)=(t-1)x(t)$. Then [19, Problem #52] $\sigma(T) = [-1,0]$, $r(T)=1$, and -1 is not an eigenvalue of T. Clearly T is bounded and self-adjoint. Now $1 \in \rho(T)$ so $I-T$

is 1-1 and onto. Theorem (1) may be applied to show that $\sum_{i=0}^{\infty} T^i y = (I-T)^{-1} y = x^*$ for any y. Note that $\|T\|=1$ and Theorem (1.6) cannot be applied. Also observe that the Neumann series converges for every y, but $r(T)=1$.

Before we prove the result of Browder and Petryshyn, we require a definition.

Definition. (5)

Let $\{A_n\}$ be a sequence of bounded linear operators on a Banach space $\mathcal{B}$. The sequence is said to converge (uniformly) to the operator A if $\lim_{n\to\infty} \|A_n - A\| = 0$. As usual this is denoted by $A_n \to A$. The sequence is said to converge strongly to A if $\lim_{n\to\infty} A_n x = Ax$ for all $x \in \mathcal{B}$. The sequence $\{A_n\}$ is said to converge in some weak sense to A if it converges pointwise to A in the same weak sense.

Theorem. (6)

Let T be a bounded linear operator on a Banach space $\mathcal{B}$ such that $\{T^n\}$ converges strongly.

(a) If $y \in R(I-T)$, then the sequence $\{x_n\}$ given by

$$x_{n+1} = Tx_n + y, \quad x_0 \text{ arbitrary}, \tag{7}$$

converges to a solution of

$$(I-T)x = y. \tag{8}$$

(b) If any subsequence $\{x_{n_j}\}$ of (7) converges, then the whole sequence $\{x_n\}$ converges.

(c) If $\mathcal{B}$ is reflexive and the sequence $\{x_n\}$ given by (7) is bounded, then $\{x_n\}$ always converges to a solution of (8).

Proof.

Let the strong limit of $\{T^n\}$ be denoted by Q. The Uniform Boundedness Principle shows that Q is a bounded linear operator. For each $\omega \in \mathcal{B}^*$:

$$\langle x,(T^*)^n\omega\rangle = \langle T^n x,\omega\rangle \to \langle Qx,\omega\rangle = \langle x,Q^*\omega\rangle .$$

Thus $\{(T^*)^n\}$ converges weakly* to Q^*. But then

$$\langle T^{2n}x,\omega\rangle = \langle T^n x,(T^*)^n\omega\rangle \to \langle Qx,Q^*\omega\rangle = \langle Q^2x,\omega\rangle ,$$

because

$$|\langle T^n x,(T^*)^n\omega\rangle - \langle Qx,Q^*\omega\rangle| \le \|(T^*)^n\omega\| \, \|T^n x-Qx\| + |\langle Qx,(T^*)^n\omega\rangle - \langle Qx,Q^*\omega\rangle| .$$

Hence $\{T^{2n}\}$ converges weakly to Q^2. Since $\{T^n\}$ converges strongly to Q, $Q^2=Q$. Also $R(Q)=N(I-T)$ so that Q is a projection of $\mathcal{B}$ onto $N(I-T)$. For if $y \in R(Q)$, say $y=Qx$, then, since $TQx = \lim_{n\to\infty} T^{n+1}x = Qx$, we have $y=Qx=TQx=Ty$. Thus $y \in N(I-T)$. On the other hand, if $y \in N(I-T)$ so that $y=Ty$, then $y = \lim_{n\to\infty} T^n y = Qy$ or $y \in R(Q)$.

(a) Suppose that $y \in R(I-T)$. Then for some $v \in \mathcal{B}$, $(I-T)v=y$. Using (7):

$$x_n = T^n x_0 + \sum_{i=0}^{n-1} T^i y = T^n x_0 + \sum_{i=0}^{n-1} T^i (I-T)v$$

$$= T^n x_0 + v - T^n v .$$

But $T^n x_0 \to Qx_0$ and $T^n v \to Qv$, so $x_n \to v - Q(v-x_0)$. Because $Qx=TQx$, we see that this is a solution of equation (8).

(b) Suppose that $x_{n_j} \to u$ for some $u \in \mathcal{B}$. Then

$$x_{n_j} = T^{n_j} x_0 + \sum_{i=0}^{n_j-1} T^i y ,$$

so that $\lim_{n_j\to\infty} \sum_{i=0}^{n_j-1} T^i y = u-Qx_0$. Since $QT=Q$, we have

$$Qx_{n_j} = QT^{n_j} x_0 + Q \sum_{i=0}^{n_j-1} T^i y = Qx_0 + n_j Qy$$

for n_j arbitrarily large. Thus $Qy=0$. This means that

$$(I-T) \sum_{i=0}^{n_j-1} T^i y = y - T^{n_j} y \to y - Qy = y .$$

Therefore $(I-T)(u-Qx_0)=y$ or $y \in R(I-T)$. Part (a) now shows that $\{x_n\}$ converges. Clearly $x_n \to u$ and u solves equation (8).

(c) Let $\mathcal{B}$ be reflexive and $\{x_n\}$ be bounded. The Uniform Boundedness Principle shows that $\|T^n\| \le M_1$. In addition, $\|x_n\| \le M_2$ by hypothesis. Thus

$$\|\sum_{i=0}^{n} T^i y\| = \|x_{n+1} - T^{n+1} x_0\| \le \|x_{n+1}\| + \|T^{n+1}\| \, \|x_0\| \le M . \tag{9}$$

Define $y_n = y - \frac{1}{n} \sum_{i=0}^{n-1} T^i y$ for $n=1,2,3,\ldots$. Inequality (9) shows that $y_n \to y$. Now define

$$u_m = \sum_{i=0}^{m-1} T^i y , \quad z_n = \frac{1}{n} \sum_{m=1}^{n-1} u_m ,$$

and observe that $\|z_n\| \le M$. Since $y_n = (I-T)z_n$, it is clear that $y_n \in R(I-T)$. Because the sequence $\{z_n\}$ is bounded, the reflexiveness of $\mathcal{B}$ insures that $\{z_n\}$ has a weakly convergent subsequence [41, p. 209]. Suppose that $\{z_{n_j}\}$ converges weakly to $z \in \mathcal{B}$. Then $\{(I-T)z_{n_j}\}$ converges weakly to $(I-T)z$. But $y_{n_j} = (I-T)z_{n_j} \to y$. Thus $y=(I-T)z$ or $y \in R(I-T)$. Part (a) now yields the result. ▌

We have already commented that if the sequence given by (7) converges, then it must converge to a solution of equation (8). Part (b) is interesting because it shows that if any subsequence $\{x_{n_j}\}$ converges, then the sequence $\{x_n\}$ converges. Thus convergence of the method of successive approximations can be tested by testing <u>any</u> subsequence. At this point we are going to prove one of the lemmas which will enable us to obtain from Theorem (6) a proof of Krasnoselskii's result, Theorem (1). First, observe that if the sequence $\{T^n\}$ converges strongly, then -1 is not an eigenvalue of T .

<u>Lemma</u>. (10)

Let T be a bounded linear operator on a Banach space $\mathcal{B}$ such that -1 is not an eigenvalue of T. If $\{T^{2n}\}$ converges strongly, then $\{T^n\}$ converges strongly also.

Proof.

Denote the strong limit of $\{T^{2n}\}$ by Q. Just as in the proof of Theorem (6), one shows that Q is a projection onto $N(I-T^2)$. Because -1 is not an eigenvalue of T, $N(I-T^2) = N[(I-T)(I+T)] = N(I-T)$. Thus $R(Q) = N(I-T)$ and $TQ=Q$. Now $\lim_{n\to\infty} T^{2n+1}x = \lim_{n\to\infty} T(T^{2n})x = TQx = Qx$. Thus $\{T^n\}$ converges strongly to Q. ▌

Before proving the Hilbert space result from which one obtains Krasnoselskii's theorem, we require a basic lemma.

Lemma. (11)

Let $\{A_n\}$ be a monotonically decreasing sequence of positive, self-adjoint, bounded linear operators on a Hilbert space $\mathcal{H}$. Then there exists a positive, self-adjoint, bounded linear operator A on $\mathcal{H}$ such that the sequence $\{A_n\}$ converges strongly to A.

Proof.

It suffices to consider the case $I \geq A_1 \geq A_2 \geq \ldots \geq 0$. For $m > n$, $A_n-A_m \geq 0$ and $\|A_n-A_m\| \leq 1$. Using the generalized Schwarz inequality:

$$\|A_nx-A_mx\|^4 \leq \langle(A_n-A_m)x,x\rangle\langle(A_n-A_m)^2x,(A_n-A_m)x\rangle$$
$$\leq (\langle A_nx,x\rangle - \langle A_mx,x\rangle)\|x\|^2, \quad x \in \mathcal{H}.$$

But $\{\langle A_n x,x\rangle\}$ is a convergent sequence of real numbers. Thus $\{A_n x\}$ is convergent for each $x \in \mathscr{H}$. Denote the limit by Ax. Clearly the operator A thus defined has the required properties. ▌

Theorem. (12)

Let T be a bounded and self-adjoint operator on a complex Hilbert space $\mathscr{H}$ such that $\|T\| \leq 1$. Then -1 is not an eigenvalue of T if and only if the sequence $\{T^n\}$ converges strongly.

Proof.

Suppose that -1 is not an eigenvalue of T. Since $0 \leq T^2 \leq I$, the sequence $\{T^{2n}\}$ converges strongly by Lemma (11). Lemma (10) now shows that $\{T^n\}$ converges strongly. We have already observed that the converse is true. ▌

One can now see that the result of Krasnoselskii given in Theorem (1) is an immediate consequence of this stronger result of Browder and Petryshyn.

The method of successive approximations often exhibits a slow rate of convergence. Just how slow will be mentioned at the end of Section 3. We now look at a technique for obtaining faster convergence. The following scheme, due to Petryshyn [123], is a corollary of the result of Browder and Petryshyn, Theorem (6). In [123] Petryshyn gives a direct proof of his result, but we shall treat it here as a corollary to Theorem (6). This method replaces 2^n steps of the usual method of successive approximations with only n steps, thus accelerating the convergence. These techniques were first applied in the field of linear algebra by Naisul [31].

Theorem. (13)

Let T be a bounded and self-adjoint operator on a complex Hilbert space $\mathcal{H}$ such that $\|T\| \leq 1$. If -1 is not an eigenvalue of T, then the sequence $\{u_n\}$ given by

$$u_1 = T^2 u_0 + (I+T)y \ , \quad u_n = T^{2^{n-1}} u_{n-1} + U_{2^{n-1}} y \ , \quad n=2,3,\ldots, \tag{14}$$

where u_0 is arbitrary and

$$U_1 = I, \ U_{2^{n-1}} = (I+T^{2^{n-2}})U_{2^{n-2}}, \quad n=2,3,\ldots, \tag{15}$$

converges to a solution of equation (2) if and only if (2) is solvable.

Proof.

We shall show by induction that $u_n = x_{2^n}$, $n \geq 1$, where the sequence $\{x_n\}$ is given by (7) with x_0 chosen equal to u_0. Because of Theorem (12), the result will then follow at once from Theorem (6).

We make use of the expression for x_{n+1} given in (1.4). For $n=1$,

$$u_1 = T^2 u_0 + (I+T)y = T^2 x_0 + \sum_{i=0}^{1} T^i y = x_2 \ .$$

Assume now that $u_{n-1} = x_{2^{n-1}}$. Then

$$u_n = T^{2^{n-1}} u_{n-1} + U_{2^{n-1}} y = T^{2^{n-1}} x_{2^{n-1}} + (I+T^{2^{n-2}})U_{2^{n-2}}y$$

$$= T^{2^{n-1}}(T^{2^{n-1}} x_0 + \sum_{i=0}^{2^{n-1}-1} T^i y) + (I+T^{2^{n-2}})(I+T^{2^{n-3}})\ldots(I+T)y$$

$$= T^{2^n} x_0 + \sum_{i=2^{n-1}}^{2^n-1} T^i y + \sum_{i=0}^{2^{n-1}-1} T^i y = x_{2^n} . \blacksquare$$

As a corollary to this result, Petryshyn obtains the following theorem which is concerned with solving an equation of the first kind using an arbitrary bounded operator. The origins of the techniques for writing the equation of the first kind as an equivalent equation of the second kind used in this theorem are due to Wiarda [46], Natanson [32], and Fridman [15] who have observed special cases. The method is very general and always converges whenever the equation is solvable without any restrictions on the operator, other than it be bounded.

Theorem. (16)

Let B be a bounded linear operator on a Hilbert space $\mathcal{H}$, and let α be such that

$$0 < \alpha < \frac{2}{\|B^*B\|} . \tag{17}$$

Let P denote the orthogonal projection of $\mathcal{H}$ onto $\overline{R(B)}$. The sequence $\{u_n\}$ given by (14) and (15), where

$$y = \alpha B^*g \ , \quad T = I - \alpha B^*B \ ,$$

converges to a solution of

$$Bx = Pg \ , \quad g \in \mathcal{H} \ , \tag{18}$$

if and only if (18) is solvable.

<u>Proof.</u>

First, note that $B^*Pg = B^*g$. For if $g = n + r$ with $n \in N(B^*)$, $r \in N(B^*)^{\perp} = \overline{R(B)}$, then $Pg = r$ shows that $B^*Pg = B^*r = B^*g$. Thus Lemma (I.1.8) shows that (18) is equivalent to $B^*Bx = B^*g$ when (18) is solvable. Multiplying by the scalar α , this may in turn be written in the form $(I-T)x = y$. To complete the proof we need only show that T is a bounded and self-adjoint operator on $\mathcal{H}$ with $\|T\| \leq 1$ such that -1 is not an eigenvalue of T . For once this is shown, we can apply Theorem (13).

First, $T^* = (I - \alpha B^*B)^* = I - \alpha(B^*B)^* = I - \alpha B^*B = T$. Condition (17) shows that $\|T\| \leq 1$. If we assume that $Tx = -x$ for some $x \neq 0$, then $(I+T)x = 0$ or $(2I - \alpha B^*B)x = 0$. This means that $2\|x\| = \alpha\|B^*Bx\| \leq \alpha\|B^*B\| \, \|x\| < 2\|x\|$ by inequality (17). This is impossible. ▌

SECTION 3

A FIXED POINT THEOREM

We shall continue to seek a solution of the linear operator equation

$$Ax = y \tag{1}$$

with $A=I-T$. However, for the moment we shall make the special choice $y=0$. In this case our solutions to (1) belong to $N(A)=N(I-T)$. Thus we shall actually be seeking fixed points of the operator T . It is not our purpose here to pursue the general subject of fixed point theorems. However, we shall examine several results of this nature in order to lead up to a theorem due to Oblomskaja, which is the main result of the section. One should note that the standard successive approximation recursion formula (1.3) takes on the form

$$x_{n+1} = Tx_n , \quad x_0 \text{ arbitrary,} \tag{2}$$

when $y=0$. In 1955, Krasnoselskii [92] modified the procedure (2) in the following theorem.

Theorem. (3)

Let B be a closed and bounded, convex subset of a uniformly convex Banach space $\mathcal{B}$, and let T be a completely continuous operator on B into itself such that

$$\|Tx-Ty\| \leq \|x-y\| , \quad x, y \in B . \tag{4}$$

Then the sequence $\{x_n\}$ given by

$$x_{n+1} = \frac{1}{2} Tx_n + \frac{1}{2} x_n , \quad x_0 \in B \text{ arbitrary,} \tag{5}$$

converges to a solution x^* of $(I-T)x=0$.

In this theorem it is not required that the operator T be linear. This explains why (4) is written in such a manner. When T is linear, condition (4) can be replaced by the condition $\|T\| \leq 1$. However, we are interested in using $Tx+y$ in place of Tx , because this choice yields equation (1). Of course the operator $\hat{T}$ given by $\hat{T}x=Tx+y$ is not linear when $y \neq 0$.

A modification of Krasnoselskii's result due to Edelstein is given in [66]. A generalization of Krasnoselskii's result which uses a scheme similar to (5), but under less restrictive conditions, is given by Kaniel [75]. In 1957, Schaefer [137] generalized Krasnoselskii's result by replacing (5) by

$$x_{n+1} = (1-\alpha)Tx_n + \alpha x_n , \quad x_0 \text{ arbitrary,} \tag{6}$$

where $0 < \alpha < 1$. Schaefer also attempted to obtain the result without the hypothesis of complete continuity. However, without this hypothesis, he did not obtain the result that $x_n \to x^*$ such that $Tx^*=x^*$, but only that

$$\lim_{n\to\infty} \|Tx_n - x_n\| = 0 . \tag{7}$$

Schaefer does show that in a Hilbert space the sequence $\{x_n\}$ converges weakly to x^* whenever T is continuous, but not necessarily completely continuous. In general, condition (7) alone does not even insure that T has a fixed point.

Example. (8)

Consider the space $C[0,1]$ of continuous real-valued functions defined on $[0,1]$ with the "sup" norm. Let B be the subset of $C[0,1]$ consisting of all functions x such that $x(0)=0$, $x(1)=1$, and $0 \leq x(t) \leq 1$. Define an operator T on B by

$$Tx(t) = tx(t) , \quad t \in [0,1] .$$

Note that T has no fixed point on B. If we use Krasnoselskii's scheme (5) with $x_0(t)=t$, we have

$$x_n(t) = \frac{(1+t)^n}{2^n} t ,$$

so that

$$|Tx_n(t) - x_n(t)| = \frac{(1+t)^n}{2^n} (t-t^2) .$$

If we differentiate $(1+t)^n(t-t^2)$, we see that $nt-nt^2+1-t-2t^2=0$ is necessary for a maximum. This yields $t_0-t_0^2 = \frac{3t_0-1}{n+2}$, where the expression (9) is maximized at t_0. Thus

$$|Tx_n(t) - x_n(t)| \leq \frac{2^n}{2^n} \cdot \frac{(3-1)}{(n+2)} = \frac{2}{n+2} .$$

Therefore $\lim_{n\to\infty} \|Tx_n-x_n\| = 0$, but T has no fixed point on B.

IV.3

It would perhaps seem logical that the next generalization in the sequence of schemes (2), (5), (6) would be to replace the fixed α in (6) by some α_n varying with n. Both Reinermann [131] and Oblomskaja [113] have attempted this sort of a generalization. We shall mention the techniques of Reinermann only briefly, but will go into detail concerning the results of Oblomskaja, since he has been able to delete the hypothesis of complete continuity.

Reinermann [131] calls his procedure a Toeplitz iteration procedure. He takes a sequence of real numbers $\{\alpha_n\}$ such that $\alpha_0=1$, $0<\alpha_n\leq 1$, and $\sum_{n=0}^{\infty}\alpha_n$ diverges. Then he defines

$$t_{nk}=\begin{cases}\alpha_k\prod_{i=k+1}^{n}(1-\alpha_i) & , \ k<n\\ \alpha_n & , \ k=n\\ 0 & , \ k>n.\end{cases} \tag{10}$$

The matrix (t_{nk}) is a Toeplitz matrix; i.e., $\sup_n \sum_{k=0}^{\infty}|t_{nk}|$ is finite, $\lim_{n\to\infty} t_{nk}=0$, and $\lim_{n\to\infty}\sum_{k=0}^{\infty} t_{nk}=1$. Reinermann's iteration scheme is of the form

$$x_{n+1}=\sum_{k=0}^{n} t_{nk}\,Tx_k\ , \quad x_0 \text{ arbitrary}, \tag{11}$$

or what is the same,

$$x_{n+1}=\alpha_n Tx_n+(1-\alpha_n)x_n\ , \quad x_0 \text{ arbitrary}. \tag{12}$$

The choice $\alpha_n=1$ for all n produces (2), while the choice $\alpha_n=\frac{1}{2}$

produces (5) and $\alpha_n = (1-\alpha)$ produces (6). We also point out that the choice $\alpha_n = \frac{1}{n+1}$ leads to $t_{nk} = \frac{1}{n+1}$, $k \le n$, and $t_{nk} = 0$, $k > n$, so that (11) takes the form

$$x_{n+1} = \frac{1}{n+1} \sum_{k=0}^{n} Tx_k , \quad x_0 \text{ arbitrary.} \tag{13}$$

This special Toeplitz matrix is a Cesàro matrix, and the resulting method (13) is the method given by Mann [105] to which we shall refer in the first paragraph of Section 4.

The theorems which Reinermann proves parallel those of Schaefer. Reinermann's first result yields the same conclusion (7) that Schaefer obtained. Because he uses α_n in place of $1-\alpha$, Reinermann's hypotheses are much more complicated. Reinermann obtains convergence of the sequence $\{x_n\}$ to a fixed point only after adding the hypothesis that T^n be completely continuous for some natural number n.

We now turn to the method given by Oblomskaja [113]. His method is motivated by the fact that Schaefer's scheme (6) can be written in the form

$$x_n = \alpha^n x_0 + n\alpha^{n-1}(1-\alpha)Tx_0 + \frac{n(n-1)}{1\cdot 2}\alpha^{n-2}(1-\alpha)^2T^2x_0 + \ldots + (1-\alpha)^nT^nx_0 . \tag{14}$$

This in turn can be written more generally as

$$x_n = \sum_{i=0}^{n} \beta_{i,n}T^i x_0 , \quad x_0 \text{ arbitrary,} \tag{15}$$

which is precisely the form of the scheme used by Oblomskaja. It will be seen that Oblomskaja's method includes both Krasnoselskii's and Schaefer's as special cases, provided the operator is linear. However, the basic scheme (2) is not included because condition (c) below on the $\beta_{i,n}$ is not satisfied. We now state and prove Oblomskaja's result. Note in particular that the operator T need not be completely continuous. We shall comment later on the necessity of the hypothesis of reflexiveness.

Theorem. (16)

Let T be a linear operator on a reflexive Banach space $\mathcal{B}$ such that $\|T^n\| \leq M$ for $n=0,1,2,\ldots$. If the sequence $\{x_n\}$ is given by (15) where the $\beta_{i,n}$ satisfy:

(a) $\beta_{i,n} \geq 0$, $i=0,1,2,\ldots,n$; $n=0,1,2,\ldots$,

(b) $\sum_{i=0}^{n} \beta_{i,n} = 1$, $n=0,1,2,\ldots$,

(c) $\lim_{n\to\infty} (\beta_{0,n} + \beta_{n,n} + \sum_{i=0}^{n-1} |\beta_{i+1,n} - \beta_{i,n}|) = 0$,

then $x_n \to x^*$ such that $Tx^*=x^*$.

Proof.

The sequence $\{x_n\}$ is bounded since for all n ,

$$\|x_n\| \leq \sum_{i=0}^{n} \beta_{i,n} M\|x_0\| = M\|x_0\| .$$

Since $\mathcal{B}$ is reflexive, the sequence $\{x_n\}$ has a weak limit point x^* [41, p. 209]. Thus some subsequence, which we shall denote in this paragraph by $\{x_n\}$ also, converges weakly to x^* . Thus the subsequence $\{(I-T)x_n\}$ converges weakly to $(I-T)x^*$. That is, for all $\omega \in \mathcal{B}^*$, $\langle x_n - Tx_n, \omega\rangle \to \langle x^* - Tx^*, \omega\rangle$. But

$$\|(I-T)x_n\| = \|\sum_{i=0}^{n} \beta_{i,n}(I-T)T^i x_0\|$$

$$= \|\beta_{0,n}x_0 + \sum_{i=0}^{n-1}(\beta_{i+1,n} - \beta_{i,n})T^{i+1} x_0 - \beta_{n,n}T^{n+1} x_0\|$$

$$\leq M\|x_0\|(\beta_{0,n} + \beta_{n,n} + \sum_{i=0}^{n-1} |\beta_{i+1,n} - \beta_{i,n}|) \to 0 \qquad (17)$$

by condition (c). Thus $|\langle x_n - Tx_n, \omega\rangle| \leq \|(I-T)x_n\| \, \|\omega\| \to 0$, so that $\langle x^* - Tx^*, \omega\rangle = 0$ for all $\omega \in \mathcal{B}^*$, or $Tx^* = x^*$.

Finally, we must show that the sequence $\{x_n\}$ converges to x^*. Write x_0 in the form $x_0 = x^* + (x_0 - x^*)$. Observe that from (6) we have

$$x_n = \sum_{i=0}^{n} \beta_{i,n}T^i x_0 = \sum_{i=0}^{n} \beta_{i,n}T^i x^* + \sum_{i=0}^{n} \beta_{i,n}T^i(x_0 - x^*)$$

$$= x^* + \sum_{i=0}^{n} \beta_{i,n}T^i(x_0 - x^*) .$$

Hence it will suffice to show that $\sum_{i=0}^{n} \beta_{i,n}T^i(x_0 - x^*) \to 0$. If we write $x_0 - x_n$ in the form

$$x_0 - x_n = (I - \sum_{i=0}^{n} \beta_{i,n}T^i)x_0$$

$$= (I-T)[(1-\beta_{0,n})I + (1-\beta_{0,n}-\beta_{1,n})T + \ldots + (1-\beta_{0,n}-\beta_{1,n}- \ldots$$

$$- \beta_{n-1,n})T^{n-1}]x_0 ,$$

we see that $x_0 - x_n \in R(I-T)$. Thus $x_0 - x^* \in \overline{R(I-T)}$ because $\overline{R(I-T)}$ is weakly closed. (Note that $\overline{R(I-T)}$ is weakly closed even when $\mathcal{B}$ is not reflexive [8, p. 422].) Therefore, it suffices to show that $\sum_{i=0}^{n} \beta_{i,n} T^i y \to 0$ for all $y \in \overline{R(I-T)}$. First of all, if $y \in R(I-T)$, then $y = u - Tu$ for some $u \in \mathcal{B}$. In this case,

$$\left\| \sum_{i=0}^{n} \beta_{i,n} T^i y \right\| = \left\| \sum_{i=0}^{n} \beta_{i,n} (I-T) T^i u \right\| \to 0$$

just as in (17). Now if $y \in \overline{R(I-T)}$, then, for any $\epsilon > 0$, we can pick $y' \in R(I-T)$ such that $\|y - y'\| < \epsilon$. Using condition (b):

$$\begin{aligned} \left\| \sum_{i=0}^{n} \beta_{i,n} T^i y \right\| &\le \left\| \sum_{i=0}^{n} \beta_{i,n} T^i y' \right\| + \left\| \sum_{i=0}^{n} \beta_{i,n} T^i (y - y') \right\| \\ &\le \left\| \sum_{i=0}^{n} \beta_{i,n} T^i y' \right\| + M\epsilon . \end{aligned}$$

Since $y' \in R(I-T)$, the first term on the right goes to zero. Thus $\sum_{i=0}^{n} \beta_{i,n} T^i y \to 0$ for any $y \in \overline{R(I-T)}$ and $x_n \to x^*$. ▌

The reader should note that the hypothesis that $\mathcal{B}$ be reflexive can be omitted if we assume that the sequence $\{x_n\}$ given by (15) has a weakly convergent subsequence.

At this point we shall give a theorem which will be required in the next section. It is called the Yosida-Kakutani Mean Ergodic Theorem [47], and is an immediate corollary to Theorem (16).

<u>Theorem.</u> (18)

Let T be a linear operator on a Banach space $\mathcal{B}$ such that

$\|T^n\| \leq M$ for $n=0,1,2,\ldots,$ and such that for any $x_0 \in \mathcal{B}$ the sequence $\{x_n\}$ given by

$$x_n = \frac{1}{n} \sum_{i=1}^{n} T^i x_0$$

contains a subsequence which converges weakly to a point $x^* \in \mathcal{B}$. Then $x_n \to x^*$ and $Tx^* = x^*$.

Proof.

In light of our comments concerning the hypothesis of reflexiveness in Theorem (16), we need only take $\beta_{0,n}=0$, $\beta_{1,n}=\beta_{2,n}=\ldots=\beta_{n,n}=\frac{1}{n}$ for all n in that theorem; for such a choice clearly satisfies conditions (a), (b), and (c). ▌

Corollary. (19)

The methods of Krasnoselskii (5) and Schaefer (6) converge under the hypotheses of Theorem (16).

Proof.

It will suffice to show that (6) belongs to the class of methods of Oblomskaja, for (5) is clearly a special case of (6). In the method (6), the sequence $\{x_n\}$ is given by (14). Clearly condition (a) is satisfied. We see that $\sum_{i=0}^{n} \beta_{i,n} = [\alpha+(1-\alpha)]^n = 1$. Thus condition (b) is also satisfied. In the expression $\beta_{i+1,n}-\beta_{i,n}$, the sign changes only once as i goes from 0 to $n-1$ [12, p. 151], say after $i=p-1$. Thus

$$\beta_{0,n} + \beta_{n,n} + \sum_{i=0}^{n-1} |\beta_{i+1,n} - \beta_{i,n}| = \beta_{0,n} + \beta_{n,n} + (\beta_{1,n} - \beta_{0,n}) + \dots$$

$$+ (\beta_{p,n} - \beta_{p-1,n}) + (\beta_{p,n} - \beta_{p+1,n}) + \dots + (\beta_{n-1,n} - \beta_{n,n})$$

$$= 2\beta_{p,n} .$$

But [12, p. 184]

$$\beta_{p,n} \approx \frac{1}{\sqrt{2\pi n\alpha(1-\alpha)}} \to 0 .$$

Therefore condition (c) is also satisfied.∎

Actually Oblomskaja shows that the method given by (12) belongs to his class of methods, provided both that $0 < \alpha_n < 1$ and that the sequence $\{\alpha_n\}$ is bounded away from both 0 and 1. We shall not prove this since the verification of condition (c) is very lengthy. It is similar to the verification of (c) which we have just given, but computationally it is more difficult.

Finally, we apply the results of Oblomskaja to obtain a theorem concerning the solution of (1).

Theorem. (20)

Let T be a linear operator on a reflexive Banach space $\mathcal{B}$ such that $\|T^n\| \leq M$ for $n=0,1,2,\dots$. If the equation

$$(I-T)x = y , \quad y \in \mathcal{B} , \tag{21}$$

has at least one solution, then the sequence $\{x_n\}$ given by

$$x_{n+1} = (1-\alpha)(Tx_n+y) + \alpha x_n \,, \quad x_0 \text{ arbitrary,} \tag{22}$$

with $0 < \alpha < 1$, converges to a solution x^* of (21).

Proof.

Let u be a solution of (21). Then from (22) with $u_n = x_n - u$, we obtain

$$u_{n+1} = (1-\alpha)Tu_n + \alpha u_n \,.$$

By Corollary (19), $u_n \to u^*$ where $Tu^* = u^*$. Thus $x_n \to u^* + u = x^*$, and x^* solves (21). ▌

If one uses the stronger result of Oblomskaja mentioned just before Theorem (20), then one can obviously replace α in (22) by α_n, provided the sequence $\{\alpha_n\}$ is bounded away from 0 and 1.

In his paper [113], Oblomskaja also obtains the following interesting result. He shows that in general his method of successive approximations as given in Theorem (16) can converge at an arbitrarily slow rate. We do not give the proof.

Theorem. (23)

Let $\mathcal{B}$ be a reflexive Banach space and let $\epsilon > 0$ be given. Then there exists a linear operator T on $\mathcal{B}$ with $\|T^n\| \leq M$ such that for some sequence $\{x_n\}$ given by (15):

$$\|x_n - x^*\| = \frac{1}{n^\epsilon} \,.$$

SECTION 4

A THEOREM OF de FIGUEIREDO AND KARLOVITZ

There exist several generalizations of the fundamental result of Browder and Petryshyn given in Section 2. For the most part, these generalizations are based on an averaging technique of Mann [105] and use various ergodic theorems. A different approach is taken by Kwon and Redheffer [100]. Dotson [62, 65] gives a modified version of Eberlein's ergodic theorem [9] and uses it to obtain one generalization. Outlaw and Groetsch [114] base their generalization on G. Birkhoff's ergodic theorem [2]. The only generalization which we shall consider here is due to de Figueiredo and Karlovitz [68] and is based on the Yosida-Kakutani Mean Ergodic Theorem given in Section 3. We choose to consider this particular generalization in detail because, as a bonus, it will allow us to strengthen the Hilbert space result of Kellogg [81] given in Section III.4.

We continue to attempt to solve the equation

$$(I-T)x = y \ , \quad y \in \mathcal{B} \ , \tag{1}$$

where $T=I-A$ is a bounded linear operator on the Banach space $\mathcal{B}$. Instead of the sequence $\{x_n\}$ given by (1.3) or (1.4) considered by Browder and Petryshyn, de Figueiredo and Karlovitz consider a sequence $\{u_n\}$ given by

$$u_{n+1} = \frac{1}{n+1} Tu_0 + \frac{n}{n+1} Tu_n + y , \quad u_0 \text{ arbitrary.} \tag{2}$$

An easy induction shows that the relationship between the two sequences is given by

$$u_n = \frac{1}{n} \sum_{i=1}^{n} x_i , \tag{3}$$

provided one chooses $u_0 = x_0$. Note that the matrix (t_{nk}) given by

$$t_{nk} = \begin{cases} \frac{1}{n+1} , & k = 0 \\ 0 , & k \neq 0, k \neq n \\ \frac{n}{n+1} , & k = n , n \neq 0 \end{cases}$$

is a Toeplitz matrix. This choice of t_{nk}, when used in the scheme (3.11) of Reinermann, yields (2). But (2) is not a special case of Reinermann's result because this choice of t_{nk} cannot arise from a sequence $\{\alpha_k\}$ as in (3.10).

Before we can state and prove the result of de Figueiredo and Karlovitz, we must define a certain class of nonlinear operators and extend the Yosida-Kakutani Mean Ergodic Theorem (3.18) to a subclass of this class.

Definition. (4)

An operator A defined on a Banach space $\mathcal{B}$ is called affine if

$$A[\alpha x+(1-\alpha)y] = \alpha Ax + (1-\alpha)Ay$$

for all $x, y \in \mathcal{B}$, $0 \le \alpha \le 1$.

The affine operators in which we shall be interested are the nonlinear operators $\hat{T}$ given by $\hat{T}x=Tx+y$, where T is a bounded linear operator on $\mathcal{B}$. Observe that $x_n = \hat{T}^n x_0$. Thus (3) can be written in the form

$$u_n = \frac{1}{n}\sum_{i=1}^{n} \hat{T}^i x_0 , \quad u_0 = x_0 . \tag{5}$$

This closely resembles the formula in the Yosida-Kakutani theorem (Theorem (3.18)) and motivates our desire to extend that result to the operator $\hat{T}$.

<u>Theorem.</u> (6)

Let $\hat{T}$ given by $\hat{T}x=Tx+y$ be an affine operator on a Banach space $\mathcal{B}$ such that

$$\|\hat{T}^n x-\hat{T}^n z\| \le M\|x-z\| , \quad x, z \in \mathcal{B} , \quad n=0,1,\ldots, \tag{7}$$

for some $M > 0$, and such that for any $u_0 \in \mathcal{B}$, the sequence $\{u_n\}$ given by (5) has a subsequence $\{u_{n_j}\}$ which converges weakly to a point $u^* \in \mathcal{B}$. If the sequence $\{\hat{T}^{n_j} u_0\}$ is bounded, then $u_n \to u^*$ and $\hat{T}u^*=u^*$.

<u>Proof.</u>

Condition (7) with $n=1$ shows that $\hat{T}$ is continuous. Thus the sequence $\{\hat{T}u_{n_j} - u_{n_j}\}$ converges weakly to $\hat{T}u^*-u^*$. But

$$\|\hat{T}u_n - u_n\| = \frac{1}{n}\|\hat{T}^{n+1}u_0 - \hat{T}u_0\| \le \frac{M}{n}\|\hat{T}^n u_0 - u_0\| . \tag{8}$$

Because the sequence $\{\hat{T}^{n_j}u_0\}$ is bounded, inequality (8) shows that $\|\hat{T}u_{n_j} - u_{n_j}\| \to 0$. Thus, just as was shown following (3.17), we have $\hat{T}u^* = u^*$.

Since $\hat{T}0 = y$, an induction yields

$$\hat{T}^n x = T^n x + \hat{T}^n 0 , \tag{9}$$

for all $n \ge 1$ and all $x \in \mathcal{B}$. The use of (7) now shows that

$$\|T^n x\| = \|\hat{T}^n x - \hat{T}^n 0\| \le M\|x\| ,$$

or $\|T^n\| \le M$ for $n \ge 1$. We shall now use Theorem (3.18) to complete the proof. First, from (9):

$$\frac{1}{n}\sum_{i=1}^{n} T^i u_0 = \frac{1}{n}\sum_{i=1}^{n} \hat{T}^i u_0 - \frac{1}{n}\sum_{i=1}^{n} \hat{T}^i 0 .$$

Since $\hat{T}u^* = u^*$, we also obtain from (9):

$$\frac{1}{n}\sum_{i=1}^{n} T^i u^* = u^* - \frac{1}{n}\sum_{i=1}^{n} \hat{T}^i 0 .$$

Subtraction of the second equation from the first gives us

$$\frac{1}{n}\sum_{i=1}^{n} T^i(u_0 - u^*) = \frac{1}{n}\sum_{i=1}^{n} \hat{T}^i u_0 - u^* . \tag{10}$$

By hypothesis, the expression on the left converges weakly to zero when n is replaced by n_j . Now Theorem (3.18) shows that the expression on the left converges strongly to zero. Thus the right side of (10) also converges strongly to zero. That is, $u_n \to u^*$. ▌

We shall now give the main result of de Figueiredo and Karlovitz. Note that the theorem only requires that $\|T^n\| \leq M$. Browder and Petryshyn require that the sequence $\{T^n\}$ converge strongly. The Uniform Boundedness Principle shows that this condition implies that $\|T^n\| \leq M$. But the operator $-I$ shows that the converse is not true.

Theorem. (11)

Let T be a linear operator defined on a Banach space $\mathcal{B}$ such that $\|T^n\| \leq M$ for $n=0,1,2,\ldots$, and let the sequence $\{u_n\}$ be given by (2).

(a) If the sequence $\{T^n\}$ converges weakly and if $y \in R(I-T)$, then $u_n \to u^*$ which solves (1).

(b) If any subsequence $\{u_{n_j}\}$ converges weakly and if $\|\sum_{i=0}^{n_j} T^i y\| \leq M_1$, then the whole sequence $\{u_n\}$ converges strongly.

(c) If $\mathcal{B}$ is reflexive and if $\|\sum_{i=1}^{n} T^i y\| \leq M_2$, then $u_n \to u^*$ which solves (1).

Proof.

As in (9):

$$\hat{T}^n x = T^n x + \hat{T}^{n-1} y = T^n x + \sum_{i=0}^{n-1} T^i y \ , \quad x \in \mathcal{B} \ . \tag{12}$$

Since $\|T^n\| \leq M$ we have that

$$\|\hat{T}^n x - \hat{T}^n z\| = \|T^n x - T^n z\| \leq M\|x-z\| \ , \quad x, z \in \mathcal{B} \ .$$

(a) By hypothesis there exists some u in $\mathcal{B}$ such that $\hat{T}u=u$. Thus (12) shows that for all n,

$$u = T^n u + \hat{T}^{n-1} y .$$

Since the sequence $\{T^n\}$ converges weakly, the sequence $\{\hat{T}^n y\}$ converges weakly. Thus the sequence $\{\hat{T}^n u_0\} = \{T^n u_0 + \hat{T}^{n-1} y\}$ converges weakly, say to u^*, and is bounded. But then the sequence $\{\frac{1}{n} \sum_{i=1}^{n} \hat{T}^i u_0\}$ also must converge weakly to u^*. By Theorem (6), $\hat{T}u^*=u^*$ and $\frac{1}{n} \sum_{i=1}^{n} \hat{T}^i u_0 = u_n \to u^*$.

(b) As we have commented, the sequence $\{u_n\}$ may be obtained from (5). Thus the sequence $\{\frac{1}{n_j} \sum_{i=1}^{n_j} \hat{T}^i u_0\}$ converges weakly to some $u^* \in \mathcal{B}$ by hypothesis. The use of (12) in conjunction with the hypotheses shows that this sequence is bounded. Theorem (6) thus gives the result. Observe that $u_n \to u^*$ which solves (1).

(c) Observe that the hypothesis $\|\sum_{i=1}^{n} T^i y\| \leq M_2$ is equivalent to requiring that the sequence $\{x_n\}$ be bounded, which coincides exactly with the hypothesis in (c) of Theorem (2.6). Just as in (b), one shows that the sequence $\{\frac{1}{n} \sum_{i=1}^{n} \hat{T}^i u_0\} = \{u_n\}$ is bounded. Since $\mathcal{B}$ is reflexive, the sequence $\{u_n\}$ has a weakly convergent subsequence [41, p. 209]. This is the case considered in (b). ∎

We are now in a position to examine the improvement of Kellogg's result (Theorem (III.4.30)) given by de Figueiredo and Karlovitz.

The proof follows at once from part (c) of Theorem (11) when one uses the results of Section III.4. Thus we again consider the situation in that section. We wish to solve the linear operator equation

$$Ax = y \,, \quad y \in \mathcal{H} \,, \tag{13}$$

where $\mathcal{H}$ is a complex Hilbert space. Let A_1 and A_2 be unbounded linear operators with nonnegative real parts (Definition (III.4.17)), and set $A=A_1+A_2$ with $D(A)=D(A_1) \cap D(A_2)$. For $\rho > 0$, define $T(B)=(\rho I-B)(\rho I+B)^{-1}$, and denote $T(A_i)$ by T_i, i=1,2. As before, let $T=T_1T_2$. We do <u>not</u> make the assumptions (a) and (b) used in Lemma (III.4.22). Recall that Lemma (III.4.27) shows that x^* solves (13) if and only if $u^*=(\rho I+A_2)x^*$ solves

$$(I-T)x = T_1y + y \,. \tag{14}$$

<u>Theorem.</u> (15)

Let $\mathcal{H}$, A_1, A_2, T_1, T_2, T, and $\rho > 0$ be as previously defined. A solution x^* of (13) exists if and only if

$$\left\| \sum_{i=0}^{n} T^i(T_1y+y) \right\| \le N \,, \quad n=0,1,2,\ldots \,. \tag{16}$$

In this case, the sequence $\{u_n\}$ given by

$$u_{n+1} = \frac{1}{n+1} Tu_0 + \frac{n}{n+1} Tu_n + T_1y + y \,,$$

$u_0 \in \mathcal{H}$ arbitrary, converges to a solution of (14) and the sequence $\{x_n\}$ given by $x_n=(\rho I+A_2)^{-1}u_n$ converges to x^*.

<u>Proof.</u>

Part (b) of Lemma (III.4.18) shows that $\|T^n\| \leq M$ with $M=1$. By part (c) of Theorem (11), a solution u^* of (14) exists and $u_n \to u^*$ when (16) holds. Note that because $\|T^n\| \leq 1$, condition (16) on the partial sums of the Neumann series is necessary for the existence of a solution of (14). In light of the comment immediately preceeding the statement of the theorem, a solution x^* of (13) exists if and only if (16) holds. By part (a) of Lemma (III.4.18), the operator $(\rho I+A_2)^{-1}$ exists and is bounded. Thus $x_n \to x^*$. ▌

Using the sequence $\{v_n\}$ given by (III.4.21), the sequence $\{V_n\}$ of Theorem (III.4.30) is given by

$$V_n = \frac{1}{n+1}(v_0 + v_1 + \ldots + v_n) .$$

One can show that the sequence $\{x_n\}$ used here may be represented by

$$x_n = \frac{1}{n}(v_1 + v_2 + \ldots + v_n) .$$

Thus the sequence $\{V_n\}$ of Kellogg is very similar to the sequence $\{x_n\}$ of de Figueiredo and Karlovitz.

CHAPTER V

GRADIENT METHODS

SECTION 1

INTRODUCTION

In this chapter we shall deal with certain iterative methods in a complex Hilbert space called gradient methods. This class of methods includes the method of steepest descent and the method of conjugate gradients which is a special case of the conjugate direction method. In his early work on the method of steepest descent, Courant [5, p. 17] tells us that the techniques of the method were used by Hadamard in a 1907 paper [18] on clamped plates. Kantorovic [77, p. 95] claims that the basic idea for this method was even known to Cauchy [4]. A more modern formulation was given by Temple [42] in 1939. The use of the steepest descent method in conjunction with operator equations first appears in the works of Kantorovic [76, 77].

An early formulation of the conjugate direction method for matrices was given by Fox, Huskey, and Wilkinson [14]. The conjugate gradient method for solving systems of linear equations was developed independently by a group directed by Hestenes at the Institute for Numerical Analysis of the National Bureau of Standards and by Stiefel [39]. A complete presentation appears in their joint paper

[24]. The extension of these methods to operator equations was first given by Hayes [71] in his dissertation while a student of Hestenes.

In Section 2 we shall be concerned with the method of steepest descent. A simple proof of the basic method, which is due to Hayes [71], will be given. Then we shall discuss a method of Petryshyn [116] called the RME method which contains the method of steepest descent as a special case. This will also give us an improved error estimate for the method of steepest descent. In addition we shall look at two closely related methods due to Bessmertnyh [53]. Finally, we shall discuss a generalization of the method of steepest descent for unbounded Kpd operators due to Petryshyn [116].

In Section 3 we examine the conjugate direction method. Particular attention will be given to the conjugate gradient method, which is a special case of the conjugate direction method. The basic convergence proofs which we shall give are largely due to Hayes [71], although certain modifications are due to Daniel [60]. Also we shall re-examine the method of Samanskii (Theorem (II.3.13)) and show how it connects with the conjugate gradient method and the method of steepest descent. This will allow us to obtain a better convergence estimate for the conjugate gradient method and to compare this method with the method of steepest descent. Finally, we shall discuss the use of the method of conjugate gradients in conjunction with unbounded operator equations.

At this point we pause to prove some introductory lemmas which form a foundation for the methods to be discussed. These lemmas were first given by Hayes, although the proof of one key lemma (Lemma (5)) was not given. Subsequently a proof of the lemma was published by Antoseiwicz and Rheinboldt [52] in the case of a real Hilbert space. In the proofs which we give here, we follow the reasoning of Antoseiwicz and Rheinboldt, but we shall generalize the results to hold in a complex Hilbert space.

Let $\mathcal{H}$ be a complex Hilbert space, and let $A: \mathcal{H} \to \mathcal{H}$ be a densely defined, positive bounded below linear operator. Theorem (I.1.7) shows that A can be extended to a self-adjoint operator with $R(A) = \mathcal{H}$. We assume that this has been done so that the equation

$$Ax = y , \quad y \in \mathcal{H} , \tag{1}$$

has a unique solution x^* for every $y \in \mathcal{H}$, possibly in the generalized sense. Recalling that $r=r(x)=y-Ax$ and $e=e(x)=x^*-x$ are respectively the residual vector and error vector, we define a functional E by

$$E(x) = \langle e,r \rangle = \langle e,Ae \rangle . \tag{2}$$

We shall call this the error functional.

Lemma. (3)

Let B be a subspace of $D(A)$ and let $x' \in D(A)$. Then $E(x') \le E(x'+z)$ for all $z \in B$ if and only if $r' = y-Ax'$ satisfies $\langle r',z\rangle = 0$ for all $z \in B$.

Proof.

Suppose $\langle r',z\rangle = 0$ for all $z \in B$. Then $E(x'+z)-E(x') = \langle z,Az\rangle-\langle r',z\rangle-\langle z,r'\rangle = \langle z,Az\rangle \ge 0$. Conversely, if $E(x') \le E(x'+z)$ for all $z \in B$, then

$$E(x') - E(x'+tz) = 2t \operatorname{Re}\langle r',z\rangle - t^2\langle z,Az\rangle \le 0$$

for all $z \in B$, t real. It follows that this expression has a maximum when $t=0$. Consideration of the parabola $s = 2\operatorname{Re}\langle r',z\rangle t - \langle z,Az\rangle t^2$ shows that $\operatorname{Re}\langle r',z\rangle = 0$. Replacing t by it in the above yields $\operatorname{Im}\langle r',z\rangle = 0$. Thus $\langle r',z\rangle = 0$ for all $z \in B$. ▌

Corollary. (4)

$E(x') \le E(x)$ for all $x \in D(A)$ if and only if $x' = x^*$.

Lemma. (5)

Let B be a closed subspace of $\mathcal{H}$ contained in $D(A)$. There exists a unique $x' \in B$ minimizing $E(x)$ on B. For this x', $\langle r',z\rangle = 0$ for all $z \in B$.

Proof.

Since A is positive, it is clear that

$$\gamma = \inf\{E(z) : z \in B\} \geq 0 .$$

By definition of γ, we may select a sequence $\{x_n\}$ in B such that $E(x_n) = \gamma_n \to \gamma$. For any real t and all $z \in B$, we have $E(x_n+tz) \geq \gamma$. Thus

$$E(x_n) - E(x_n+tz) \leq \gamma_n - \gamma , \tag{6}$$

which may be written in the form

$$2t \operatorname{Re}\langle r_n,z\rangle - t^2\langle z,Az\rangle \leq \gamma_n - \gamma .$$

The choice $t = \dfrac{\operatorname{Re}\langle r_n,z\rangle}{\langle z,Az\rangle}$ yields

$$[\operatorname{Re}\langle r_n,z\rangle]^2 \leq (\gamma_n-\gamma)\langle z,Az\rangle . \tag{7}$$

Inequality (7) shows that both $\operatorname{Re}\langle A(x^*-x_m),z\rangle \leq [(\gamma_m-\gamma)\langle z,Az\rangle]^{1/2}$ and $-\operatorname{Re}\langle A(x^*-x_n),z\rangle \leq [(\gamma_n-\gamma)\langle z,Az\rangle]^{1/2}$. Letting $z=x_n-x_m$ with $x_n \neq x_m$ and adding these two inequalities, we obtain

$$\langle z,Az\rangle \leq [(\gamma_m-\gamma)\langle z,Az\rangle]^{1/2} + [(\gamma_n-\gamma)\langle z,Az\rangle]^{1/2} .$$

Division by $\langle z,Az\rangle^{1/2}$ yields

$$(\gamma_m-\gamma)^{1/2} + (\gamma_n-\gamma)^{1/2} \geq \langle x_n-x_m,A(x_n-x_m)\rangle^{1/2} .$$

Because A is positive bounded below, this inequality shows that $\{x_n\}$ is a Cauchy sequence. Since $\mathscr{V}$ is complete and B is closed in $\mathscr{V}$, there exists $x' \in B$ such that $x_n \to x'$.

Replacing t by it in (6) and proceeding as above, we obtain

$$2t \operatorname{Im}\langle r_n,z\rangle - t^2\langle z,Az\rangle \leq \gamma_n-\gamma ,$$

and the choice $t = \frac{\text{Im}\langle r_n, z\rangle}{\langle z, Az\rangle}$ yields

$$[\text{Im}\langle r_n, z\rangle]^2 \le (\gamma_n - \gamma)\langle z, Az\rangle . \tag{8}$$

Combining (7) and (8), we get

$$|\langle r_n, z\rangle|^2 \le 2(\gamma_n - \gamma)\langle z, Az\rangle . \tag{9}$$

Now for any $z \in B$, using (9):

$$\begin{aligned} |\langle r', z\rangle| &= |\langle y - Ax', z\rangle| = |\langle A(x^* - x'), z\rangle| \\ &\le |\langle A(x^* - x_n), z\rangle| + |\langle A(x_n - x'), z\rangle| \\ &\le \{|\langle r_n, z\rangle| + \|Az\| \, \|x_n - x'\|\} \to 0 . \end{aligned}$$

Thus $\langle r', z\rangle = 0$ for all $z \in B$. By Lemma (3), $E(x') \le E(z)$ for all $z \in B$. Hence $E(x') = \gamma$.

Finally suppose that $x'' \in B$ also minimizes $E(x)$ on B. Then, by Lemma (3), both $\langle A(x^* - x'), z\rangle = 0$ and $\langle A(x^* - x''), z\rangle = 0$ for all $z \in B$. Thus $\langle A(x' - x''), z\rangle = 0$ for all $z \in B$. But $x' - x'' \in B$. Hence the choice $z = x' - x''$, along with the positive definiteness of A, shows that $x' = x''$ or x' is unique. ▌

<u>Lemma.</u> (10)

Let $B_1 \subset B_2 \subset B_3 \subset \ldots$ be an expanding sequence of closed subspaces of $\mathcal{H}$, contained in $D(A)$, and let $\{x_n\}$ be the sequence defined by $x_n \in B_n$ and $E(x_n) = \inf\{E(x) : x \in B_n\}$. Then $x_n \to x' \in \mathcal{H}$. If A is bounded, then $\langle r', z\rangle = 0$ for all $z \in \overline{\bigcup_{n=1}^{\infty} B_n}$. If $\bigcup_{n=1}^{\infty} B_n = D(A)$, then $x' = x^*$.

Proof.

Lemma (5) shows that the sequence $\{x_n\}$ is well-defined. Since $E(x_1) \geq E(x_2) \geq \ldots \geq 0$, there is a $\gamma \geq 0$ such that $E(x_n) \to \gamma$. In the same manner as in Lemma (5), one shows that $\{x_n\}$ is a Cauchy sequence. Thus $x_n \to x' \in \mathcal{H}$. Given any $z \in \bigcup_{n=1}^{\infty} B_n \subset D(A)$, there is a smallest integer $n_0 = n_0(z)$ such that $z \in B_{n_0}$. Lemma (5) asserts that $\langle r_n, z\rangle = 0$ for $n \geq n_0$. Thus $\langle A(x^*-x_n), z\rangle = 0$ or $\langle x^*-x_n, Az\rangle = 0$ for $n \geq n_0$. Thus

$$\langle x^*-x', Az\rangle = 0 \quad \text{for all} \quad z \in \bigcup_{n=1}^{\infty} B_n .$$

If A is bounded, then $\langle A(x^*-x'), z\rangle = \langle r', z\rangle = 0$ for all $z \in \overline{\bigcup_{n=1}^{\infty} B_n}$. If $\bigcup_{n=1}^{\infty} B_n = D(A)$, then, because A is self-adjoint, we see that $x^*-x' \in D(A^*) = D(A)$. By putting $z = x^*-x'$ we have $\langle x^*-x', A(x^*-x')\rangle = 0$ or $x' = x^*$. ▌

Let $p \in D(A)$, $p \neq 0$, be a vector having arbitrary direction, and let $\alpha = \frac{\langle r,p\rangle}{\langle p,Ap\rangle}$. Then using $\tau + \bar{\tau} - |\tau|^2 = 1 - |1-\tau|^2$ for $\tau \in \mathbb{C}$, we obtain

$$E(x) - E(x+\tau\alpha p) = [1-|1-\tau|^2] \frac{|\langle r,p\rangle|^2}{\langle p,Ap\rangle} . \tag{11}$$

For $\tau \in \mathbb{C}$ fixed, it can be shown that this difference is a maximum in absolute value when p has the direction of the residual vector $r = r(x)$. That is, the gradient of the functional E is in the direction of the residual vector. For this reason, r is often said to have the direction of steepest descent for E at x.

In the gradient methods we shall minimize $E(x)$ successively in the direction of the linearly independent vectors $p_0, p_1, p_2, \dots$ which are to be determined. Thus we consider an iterative scheme of the form

$$x_{n+1} = x_n + \alpha_n p_n , \quad \alpha_n = \frac{\langle r_n, p_n \rangle}{\langle p_n, Ap_n \rangle} . \tag{12}$$

Note that the value $\tau=1$ in (11) leads to a maximum difference. Thus $E(x_n)-E(x_{n+1}) = E(x_n)-E(x_n+\alpha_n p_n)$ is maximized with our choice of α_n. All our gradient methods will differ only in the choice of the sequence $\{p_n\}$. In the method of steepest descent, we select $p_n=r_n$ since this is the direction of maximum change for $E(x)$ at x_n. In the conjugate direction method, we select the p_n so that $\langle p_n, Ap_m \rangle = 0$ for $n \neq m$. Such a collection $\{p_n\}$ is called a collection of conjugate (A-conjugate) directions. In the conjugate gradient method, the p_n are chosen iteratively using the formula

$$p_{n+1} = r_{n+1} + b_n p_n , \quad b_n = \frac{-\langle r_{n+1}, Ap_n \rangle}{\langle p_n, Ap_n \rangle} .$$

In Section 3 we shall show that this is indeed a conjugate direction method.

In his dissertation [60], Daniel is concerned mainly with the conjugate gradient method. However, it is still necessary for him to consider the introductory lemmas which we have just given. He treats the situation in a real Hilbert space also following Antoseiwicz and Rheinboldt. But Daniel's operator A is somewhat more general. He takes A to be 1-1 and onto H . Then he selects an operator B which is positive bounded below and self-adjoint, so that $N=A^*BA$ is also positive bounded below and self-adjoint. He defines the functional $E(x)$ by $E(x) = \langle e,Ne\rangle = \langle r,Br\rangle$ and proceeds to prove appropriately modified versions of Lemma (3), Lemma (5), and Lemma (10). We do not give the details in this paper.

SECTION 2

THE METHOD OF STEEPEST DESCENT

We shall not engage in a discussion of the early formulations of the method of steepest descent. Kantorovic [76, 77] gave the first convergence proof in the context of linear operator equations. We do not discuss his proof, but we shall mention his geometric interpretation of the method [77] in order to gain some intuitive insight into the method of steepest descent. The proof which we give of the basic method is due to Hayes [71]. We shall also give Hayes' simple error estimate, although it is not so good as the error estimate derived by Kantorovic in [77] using spectral theory. We shall give a proof of an iterative scheme of Petryshyn [116] called the method of relative minimal errors (RME method), and as a corollary obtain the method of steepest descent and an error estimate which is at least as good as that of Kantorovic. The derivation of this estimate does not require spectral theory. Next we shall discuss a variation of the method of steepest descent which is due to Bessmertnyh [53]. Finally, we shall give a generalization of the method of steepest descent for unbounded Kpd operators. This generalization is due to Petryshyn [116].

Before proceeding we wish to mention certain results which are not discussed in detail. In [95] an iterative scheme called an α-process is discussed. This has the form

$$x_{n+1} = x_n + \frac{\langle r_n, A^\alpha r_n \rangle}{\langle A^{\alpha+1} r_n, r_n \rangle} r_n .$$

The case $\alpha=0$ is the method of steepest descent. Birman [54] also describes a method which has the method of steepest descent as a special case. He uses the iteration formula

$$x_{n+1} = x_n + \sum_{i=0}^{p-1} \alpha_{i,n} A^i r_n$$

which reduces to $x_{n+1} = x_n + \alpha_n r_n$ for $p=1$. His error estimate coincides with that of Kantorovic in case $p=1$.

A different approach to the method of steepest descent is given by Rosenbloom in [132]. In this article Rosenbloom considers a path of steepest descent which is a smooth curve as opposed to the straight line polygonal path which we use here. A steepest descent method for bounded, self-adjoint operators is given by Fridman in [69]. His work improves Kantorovic's in that he shows that the method converges to a solution whenever one exists. Nashed [110] considers operator equations which do not necessarily have a solution. He shows that the method of steepest descent converges to the least square solution of minimal norm, and he relates this "solution" to the generalized inverse of the operator. He requires that the operator have a closed range. However, this restriction is removed in a joint paper by Kammerer and Nashed [72]. Since we are not considering operator equations which are not solvable or generalized inverses of operators, we shall not discuss these interesting papers in detail.

Now we give the convergence proof of Hayes [71] for the basic method of steepest descent. We wish to solve the linear operator equation

$$Ax = y \ , \quad y \in \mathscr{Y} \ . \tag{1}$$

We use the iterative scheme (1.12), and the p_n vectors are selected at each step to be equal to r_n, the nth residual.

Theorem. (2)

Let the operator A be bounded and positive bounded below. The sequence $\{x_n\}$ given by

$$x_{n+1} = x_n + \alpha_n r_n \ , \quad x_0 \ \text{arbitrary,} \tag{3}$$

where

$$\alpha_n = \frac{\langle r_n, r_n \rangle}{\langle r_n, Ar_n \rangle} \ , \tag{4}$$

converges to the unique solution x^* of (1).

Proof.

Equation (1.11) with $\tau=1$ shows that

$$E(x_n) - E(x_{n+1}) = E(x_n) - E(x_n + \alpha_n p_n) = \frac{\|r_n\|^4}{\langle r_n, Ar_n \rangle} > 0 \ .$$

Thus the sequence $\{E(x_n)\}$ is monotonically decreasing. Being bounded below by zero, it must converge. Since

$\frac{\|r_n\|^4}{\langle r_n, Ar_n\rangle} \geq \frac{\|r_n\|^2}{\|A\|}$, we see that $r_n \to 0$ or $Ax_n \to y$. But $\|x_n - x^*\| \leq \|A^{-1}\| \, \|Ax_n - y\|$ so $x_n \to x^*$. ∎

Kantorovic [77] gives the following intuitive description of the method of steepest descent. The equations $E(x)=c$, c a constant, describe a family of similar infinite-dimensional ellipsoids with centers at the point where the functional attains its minimum value, namely at x^* . We start at the arbitrary point x_0 which is located on some ellipsoid. We proceed in the direction of the gradient (along the inward normal to the ellipsoid at x_0) to a point at which $E(x)$ assumes its minimum value. This is the point at which the normal is tangent to yet another ellipsoid of the family and is the point x_1 . Note that this new ellipsoid must be interior to the original one and that x_1 is closer to x^* than is x_0 . Also note that as we repeat the procedure, now starting from x_1 , that we must go in a direction normal to the previous direction, for the previous direction is tangent to the ellipsoid at x_1 , and we shall proceed to x_2 along the normal to the ellipsoid at x_1 .

Theorem. (5)

The sequence $\{x_n\}$ determined by the method of steepest descent converges at least as fast as a geometric progression with ratio

$$\left(\frac{M-m}{M}\right)^{1/2} , \tag{6}$$

where $0 < mI \leq A \leq MI$.

Proof.

Define

$$\rho_n = \frac{\langle r_n, A^{-1} r_n \rangle}{\langle r_n, r_n \rangle} , \quad \mu_n = \frac{\langle r_n, A r_n \rangle}{\langle r_n, r_n \rangle} .$$

Then, from (1.11),

$$E(x_n) - E(x_{n+1}) = E(x_n) - E(x_n + \alpha_n r_n) = \frac{\langle r_n, r_n \rangle}{\mu_n} . \tag{7}$$

But

$$E(x_n) = \langle e_n, A e_n \rangle = \langle A^{-1} r_n, r_n \rangle = \langle r_n, A^{-1} r_n \rangle = \rho_n \langle r_n, r_n \rangle .$$

Thus, substituting for $\langle r_n, r_n \rangle$ in (7):

$$E(x_n) - E(x_{n+1}) = \frac{E(x_n)}{\rho_n \mu_n} ,$$

or

$$E(x_{n+1}) = \left(1 - \frac{1}{\rho_n \mu_n}\right) E(x_n) . \tag{8}$$

But $\rho_n \mu_n \leq \|A\| \, \|A^{-1}\| \leq \frac{M}{m}$, so that

$$E(x_{n+1}) \leq \left(\frac{M-m}{M}\right) E(x_n) . \tag{9}$$

Finally

$$E(x_n) = \langle x^* - x_n, A(x^* - x_n) \rangle \geq m \|x^* - x_n\|^2 \tag{10}$$

so that $\{x_n\}$ converges at the asserted rate. ▌

Corollary. (11)

The sequences $\{E(x_n)\}$ and $\{r_n\}$ determined by the method of steepest descent converge at least as fast as geometric progressions with ratios $\frac{M-m}{M}$ and $\left(\frac{M-m}{M}\right)^{1/2}$ respectively.

Proof.

Inequality (9) showed the claim concerning $\{E(x_n)\}$. Since

$$E(x_n) = \langle r_n, A^{-1}r_n\rangle \geq \frac{1}{M}\|r_n\|^2 ,$$

the claim concerning $\{r_n\}$ follows also. ▌

This error estimate is a modification of a result due to Hayes [71]. Kantorovic obtains the ratio $\frac{M-m}{M+m}$ which is better than (6), although his proof is much more complicated than the proof here. If we use Greub and Rheinboldt's generalization of Kantorovic's inequality (II.2.8), we have

$$\frac{1}{\rho_n\mu_n} = \frac{\langle r_n, r_n\rangle^2}{\langle r_n, A^{-1}r_n\rangle\langle r_n, Ar_n\rangle} \geq \frac{4mM}{(m+M)^2} .$$

Thus (8) becomes

$$E(x_{n+1}) \leq \left(1 - \frac{4mM}{(m+M)^2}\right) E(x_n) ,$$

or

$$E(x_{n+1}) \leq \left(\frac{M-m}{M+m}\right)^2 E(x_n) .$$

Now (10) shows that Kantorovic's estimate holds. Note that

$$\left(\frac{M-m}{M}\right)^{1/2} > \frac{M-m}{M} > \frac{M-m}{M+m} , \quad m \neq M .$$

Next we shall discuss a method of Petryshyn which will allow us to obtain an estimate which is at least as good as Kantorovic's estimate.

In his dissertation [116], Petryshyn has developed an iterative method for solving (1), which he calls the method with relative minimal errors (RME method). There are several special cases of interest connected with this method, one of which is the method of steepest descent. We consider a bounded operator A which is K^*pd, K^* being the adjoint of a continuously invertible operator K. Theorem (I.1.7) assures us that (1) has a unique solution x^* for every $y \in \mathcal{H}$. We also assume that we can find a bounded linear operator B satisfying

$$\langle KAu, BAu\rangle \geq d\langle KAu, u\rangle , \quad d > 0 . \tag{12}$$

We shall work in the space $F(A,K^*)$, which is Petryshyn's generalization of the Friedrichs extension $F(A)$ explained in Chapter I. Thus the inner product is given by

$$\langle u,v\rangle_{K^*} = \langle Au, K^*v\rangle = \langle KAu, v\rangle$$

with the customary corresponding norm $\|\cdot\|_{K^*}$. Let $D = \|BA\|_{K^*}$.

From (12) we now see that in $F(A,K^*)$ we have $0 < dI \leq BA \leq DI$.

Let $x_0 \in \mathcal{H}$ be arbitrary, and define a sequence $\{x_n\}$ by

$$x_{n+1} = x_n + t_n Br_n , \tag{13}$$

where

$$t_n = \frac{\langle e_n, Br_n \rangle_{K^*}}{\langle Br_n, Br_n \rangle_{K^*}} = \frac{\langle Kr_n, Br_n \rangle}{\langle KABr_n, Br_n \rangle} , \quad r_n \neq 0 . \tag{14}$$

We note that $e_{n+1} = e_n - t_n Br_n$. The scalar t_n is chosen so that e_{n+1} is orthogonal to Br_n in the space $F(A,K^*)$; i.e., so that $\langle e_{n+1}, Br_n \rangle_{K^*} = \langle e_n - t_n Br_n, Br_n \rangle_{K^*} = 0$. Thus $t_n Br_n$ is the orthogonal projection in $F(A,K^*)$ of e_n onto Br_n. This means e_{n+1} is obtained from e_n by subtracting from it the projection of e_n onto Br_n. Hence $\|e_{n+1}\|_{K^*}$ assumes the minimum possible value for this choice of t_n. This explains Petryshyn's choice of the term "relative minimal error".

We note in passing that if $t_n = 0$, then $x_n = x_{n+1} = x_{n+2} = \cdots$ and $r_n = r_{n+1} = r_{n+2} = \ldots$. In this case continued iteration is worthless. Thus we would hope that if $t_n = 0$ we would have $r_n = 0$ also. Fortunately this is true. For using (12) and (14):

$$t_n = \frac{\langle Kr_n, Br_n \rangle}{\langle KABr_n, Br_n \rangle} = \frac{\langle KAe_n, BAe_n \rangle}{\langle KABr_n, Br_n \rangle}$$

$$\geq \frac{d\langle KAe_n, e_n \rangle}{\langle KABr_n, Br_n \rangle} = \frac{d\langle Ae_n, K^*e_n \rangle}{\langle KABr_n, Br_n \rangle}$$

$$\geq \frac{\gamma\langle e_n, e_n \rangle}{\langle KABr_n, Br_n \rangle} \geq 0$$

for some $\gamma > 0$. Thus $t_n=0$ implies $e_n=0$. But then $r_n=0$.

Theorem. (15)

Let the operator A be bounded and K*pd, and let the bounded operator B satisfy (12). The sequence $\{x_n\}$ generated by the RME method of (13) and (14) converges in $F(A,K^*)$ and in $\mathcal{H}$ to the unique solution x^* of (1) at least as fast as a geometric progression with ratio

$$\frac{D-d}{D+d} \tag{16}$$

for any choice of initial approximation x_0 .

Proof.

Observing that t_n is real, we use $e_{n+1}= e_n-t_nBr_n$ to obtain

$$\|e_{n+1}\|^2_{K*} = \|e_n\|^2_{K*} - \frac{\langle e_n,Br_n\rangle^2_{K*}}{\langle Br_n,Br_n\rangle_{K*}} .$$

Thus

$$\|e_{n+1}\|_{K*} = \left(1 - \frac{\langle e_n,BAe_n\rangle^2_{K*}}{\langle BAe_n,BAe_n\rangle_{K*}\langle e_n,e_n\rangle_{K*}}\right)^{1/2} \|e_n\|_{K*}$$

$$= \left(1 - \frac{\langle KAe_n,BAe_n\rangle^2}{\langle KABAe_n,BAe_n\rangle\langle KAe_n,e_n\rangle}\right)^{1/2} \|e_n\|_{K*}$$

$$\leq \left(1 - \frac{4dD}{(D+d)^2}\right)^{1/2} \|e_n\|_{K*}$$

$$= \left(\frac{D-d}{D+d}\right) \|e_n\|_{K*} ,$$

by Theorem (II.2.17). Thus $\|e_n\|_{K*} \to 0$ monotonically and $x_n \to x^*$ in $F(A,K^*)$. Since for some $\gamma > 0$,

$$\|u\|_{K*}^2 = \langle Au, K^*u\rangle \geq \gamma\|u\|^2 ,$$

we see that $x_n \to x^*$ in $\mathcal{H}$ also. ∎

Although the proof appears quite simple, it is based on the inequality (II.2.18), the proof of which is somewhat lengthy. In fact, Petryshyn developed the inequality specifically for use in the proof of this theorem. Since Petryshyn's inequality is a generalization of an inequality of Kantorovic [77], which Kantorovic used to obtain an estimate of the rate of convergence of the method of steepest descent, we might expect a close relationship between the ratio (16) and the ratio $\frac{M-m}{M+m}$ of Kantorovic. We now show that such is the case.

The RME method is not well-defined until specific choices have been made for the operators K and B . The most obvious choice K=B=I reduces the RME method to the method of steepest descent. In this case, A is positive bounded below, say $0 < mI \leq A \leq MI$. Note that the space $F(A,K^*)$ reduces to $F(A)$ and the norm becomes $\|\cdot\|_A$; i.e., the inner product is given by

$$\langle u,v\rangle_A = \langle Au,v\rangle .$$

Since BA=A we have $0 < dI \leq A \leq DI$ in $F(A)$. Thus Petryshyn has shown that the method of steepest descent converges at least as fast as a geometric progression with ratio $\frac{D-d}{D+d}$, while Kantorovic has shown that it converges at least as fast as a geometric progression with ratio $\frac{M-m}{M+m}$. Hence the following corollary to Theorem (15) shows that Petryshyn's estimate is at least as good as that of Kantorovic.

Corollary. (17)

With A, d, D, m, and M as above:

$$\frac{D-d}{D+d} \leq \frac{M-m}{M+m} .$$

Proof.

It follows directly (or by using Reid's inequality (II.2.13) with $x=Au$ and $K=I$) that

$$\langle A^2u,Au\rangle \leq \|A\| \langle Au,Au\rangle .$$

Now using Reid's inequality with $x=A^{1/2}u$ and $K=I$, we obtain:

$$\langle A^2u,u\rangle \leq \|A\| \langle Au,u\rangle .$$

Thus $\langle A^2u,Au\rangle \leq \|A\|^2\langle Au,u\rangle$ or $\|Au\|_A \leq \|A\| \|u\|_A$ so that $\|A\|_A \leq \|A\|$. Hence $D \leq M$. On the other hand, the use of $u=A^{-1/2}x$ in (12) with $K=B=I$ yields $\langle Ax,x\rangle \geq d\langle x,x\rangle$. Thus $m \leq d$. We have shown that

$$\frac{M}{m} \geq \frac{D}{d} \geq 1 .$$

Examination of the hyperbola given by $g(t) = \frac{t-1}{t+1}$ shows that g is monotonically increasing in the interval $[1,\infty)$. Thus

$$\frac{\frac{D}{d} - 1}{\frac{D}{d} + 1} \le \frac{\frac{M}{m} - 1}{\frac{M}{m} + 1} ,$$

which is equivalent to the desired inequality. ▌

Petryshyn makes several other interesting choices for K and B in the RME method. As one choice he lets $K=B=A^*$. This yields a method which is similar to applying the method of steepest descent to the equation $A^*Ax=A^*y$. Lemma (I.1.8) shows that (1) is equivalent to this equation whenever (1) is solvable. But $(Au,K^*u) \geq \alpha\|u\|^2$ becomes $\|Au\|^2 \geq \alpha\|u\|^2$ in this case so that A^{-1} exists on $R(A)$. Therefore the equation $A^*Ax=A^*y$ is uniquely solvable whenever $y \in R(A)$. Hence this method permits us to apply the method of steepest descent to operators which need not be self-adjoint.

In yet another choice, Petryshyn lets $K=A$ and $B=I$. Finally he lets $K=A^{-1}$ and $B=A^*$. In this case, equation (13) has the form

$$x_{n+1} = x_n + \frac{\|r_n\|^2}{\|A^*r_n\|^2} A^*r_n . \tag{18}$$

This is very close to the scheme used by Altman in Theorem (II.3.2). It brings to mind the questions: For what values of γ does the iterative scheme

$$x_{n+1} = x_n + \frac{\|r_n\|^2}{\gamma\|A^*r_n\|^2} A^*r_n$$

converge? What is the optimal choice of γ for most rapid convergence? In [116] Petryshyn devotes an entire section to comparing his various special cases of the RME method. We shall not discuss these special cases further here. The choice $K=A^{-1}$ and $B=I$, which is not mentioned by Petryshyn, yields the α-process of Krasnoselskii [95] with $\alpha = -1$. Krasnoselskii calls this the method of minimal errors.

As we mentioned in Section II.2, Bessmertnyh [53] has developed a method for real Hilbert spaces which is similar to the method of steepest descent. We shall discuss this method here in the context of a complex Hilbert space. Let the operator A be bounded and positive bounded below. Denote the lower and upper bounds of $\sigma(A)$ by m and M as usual. With x_0 arbitrary, we generate the sequence $\{x_n\}$ by the recursion formula

$$x_{n+1} = x_n + \frac{c}{a_n + cb_n} r_n , \tag{19}$$

where

$$b_n = \frac{\langle Ar_n, r_n \rangle}{\langle r_n, r_n \rangle} , \quad a_n = \frac{\langle Av_n, v_n \rangle}{\langle v_n, v_n \rangle} , \quad v_n = Ar_n - b_n r_n .$$

With the given conditions on A, equation (1) has a unique solution x^* for every $y \in \mathcal{H}$. If $r_n = 0$ for some n, then $x_n = x^*$, and we need not compute x_{n+1}. Similarly if $v_n = 0$ and $r_n \neq 0$ for some n, then $Ar_n = b_n r_n$, or r_n is an eigenvector of the operator A. When this happens $x^* = x_n + \frac{1}{b_n} r_n$, and we are done. Mainly we shall be interested in the case in which these two events fail to occur. Then $\{a_n\}$ and $\{b_n\}$ are well-defined sequences of real numbers.

One should note that the estimates of the rate of convergence given by Bessmertnyh for his two methods are not so good as the standard estimate $\frac{M-m}{M+m}$ of Kantorovic associated with the method of steepest descent. This does not mean that Bessmertnyh's method necessarily converges more slowly. But these estimates should certainly be kept in mind in view of the extra calculations involved in Bessmertnyh's methods. Although we do not go into detail, Bessmertnyh has shown that under certain conditions the sequences $\{a_n\}$ and $\{b_n\}$ converge to eigenvalues of A. If these are desired, then the additional calculations may be justified.

Bessmertnyh proves that the sequence given by (19) converges for $c=1$ and $c=2$, leaving open the question of exactly which values for c yield convergence. We shall give a proof for the case $c=2$. First we require some lemmas. The first lemma is due to Krasnoselskii [26]. It is often called the moment inequality.

<u>Lemma.</u> (20)

Let A be a bounded and positive bounded below operator. Let s be any real number, and let p and r be positive real numbers. Then for all $x \in \mathcal{H}$,

$$\langle A^s x,x\rangle^{p+r} \leq \langle A^{s+r}x,x\rangle^p \langle A^{s-p}x,x\rangle^r . \tag{21}$$

Proof.

Use the spectral theorem with $A = \int_m^M \lambda dE_\lambda$ to write $\langle A^s x,x\rangle = \int_m^M \lambda^s d\langle E_\lambda x,x\rangle$. We can express λ^s in the form

$$\lambda^s = \lambda^{\frac{p(s+r)}{p+r}} \lambda^{\frac{r(s-p)}{p+r}} .$$

Let $p_1 > 1$, $q_1 > 1$, and $\frac{1}{p_1} + \frac{1}{q_1} = 1$, and apply Hölder's inequality:

$$\langle A^s x,x\rangle \leq \left(\int_m^M \lambda^{\frac{p(s+r)p_1}{p+r}} d\langle E_\lambda x,x\rangle\right)^{1/p_1} \left(\int_m^M \lambda^{\frac{r(s-p)q_1}{p+r}} d\langle E_\lambda x,x\rangle\right)^{1/q_1} .$$

Now set $p_1 = \frac{p+r}{p}$ and $q_1 = \frac{p+r}{r}$ in order to obtain

$$\langle A^s x,x\rangle \leq \left(\int_m^M \lambda^{s+r} d\langle E_\lambda x,x\rangle\right)^{\frac{p}{p+r}} \left(\int_m^M \lambda^{s-p} d\langle E_\lambda x,x\rangle\right)^{\frac{r}{p+r}} ,$$

which is equivalent to (21).▌

Krasnoselskii proves the lemma for A unbounded and positive definite. The reader should observe that the proof is easily modified to hold in this case. However, we do not require this stronger result here.

Lemma. (22)

With A, m, and M as above:

$$\frac{\langle A^2x,x\rangle^3}{\langle A^3x,x\rangle^2\langle x,x\rangle} \geq \frac{64m^3M^3}{(M+m)^6} , \tag{23}$$

and

$$\frac{\langle A^2x,x\rangle^2}{\langle A^3x,x\rangle} \leq \langle Ax,x\rangle \tag{24}$$

for all $x \in \mathcal{H}$, $x \neq 0$.

Proof.

From inequality (II.2.8) we have

$$\frac{\langle x,x\rangle^2}{\langle Ax,x\rangle\langle A^{-1}x,x\rangle} \geq \frac{4Mm}{(M+m)^2} .$$

Replacing x by $A^{1/2}x$ and then by Ax , we obtain

$$\frac{\langle Ax,x\rangle^2}{\langle A^2x,x\rangle\langle x,x\rangle} \geq \frac{4Mm}{(M+m)^2} \tag{25}$$

and

$$\frac{\langle A^2x,x\rangle^2}{\langle A^3x,x\rangle\langle Ax,x\rangle} \geq \frac{4Mm}{(M+m)^2} . \tag{26}$$

Squaring (26) and multiplying it by (25), we obtain (23). Lemma (20) with $s=2$, $p=r=1$ gives (24). ▌

Theorem. (27)

Let A be a bounded and positive bounded below operator on a complex Hilbert space $\mathcal{H}$ with $0 < mI \leq A \leq MI$. The sequence $\{x_n\}$ given by (19) with $c=2$ converges to x^* at least as fast as a geometric progression with ratio

$$\left(1 - \frac{256m^4M^3}{(M+m)^5(M+2m)^2}\right)^{1/2}$$

for any choice of $x_0 \in \mathcal{H}$.

Proof.

We first note that

$$e_{n+1} = e_n - \frac{2}{a_n + 2b_n} Ae_n .$$

Thus

$$\|e_{n+1}\|^2 = \|e_n\|^2 - \frac{4}{a_n+2b_n} \langle Ae_n,e_n\rangle + \frac{4}{(a_n+2b_n)^2} \langle A^2e_n,e_n\rangle$$

$$= \|e_n\|^2 - \frac{4}{a_n+2b_n} \langle Ae_n,e_n\rangle + \frac{4}{(a_n+2b_n)b_n} \langle A^2e_n,e_n\rangle$$

$$- \frac{4}{(a_n+2b_n)b_n} \langle A^2e_n,e_n\rangle + \frac{4}{(a_n+2b_n)^2} \langle A^2e_n,e_n\rangle . \qquad (28)$$

Since

$$b_n = \frac{\langle Ar_n,r_n\rangle}{\langle r_n,r_n\rangle} = \frac{\langle A^3e_n,e_n\rangle}{\langle A^2e_n,e_n\rangle} ,$$

we have from (24) that

$$- \frac{4}{a_n+2b_n} \langle Ae_n,e_n\rangle + \frac{4}{(a_n+2b_n)b_n} \langle A^2e_n,e_n\rangle$$

$$= \frac{4}{a_n+2b_n} \left(- \langle Ae_n,e_n\rangle + \frac{\langle A^2e_n,e_n\rangle^2}{\langle A^3e_n,e_n\rangle}\right) \leq 0 .$$

Hence (28) becomes

$$\|e_{n+1}\|^2 \le \|e_n\|^2 - \frac{4}{(a_n+2b_n)b_n}\langle A^2e_n,e_n\rangle + \frac{4}{(a_n+2b_n)^2}\langle A^2e_n,e_n\rangle$$

$$= \left(1 - \frac{4(a_n+b_n)\langle A^2e_n,e_n\rangle}{(a_n+2b_n)^2 b_n\langle e_n,e_n\rangle}\right)\|e_n\|^2 . \tag{29}$$

Using (23):

$$\frac{4(a_n+b_n)\langle A^2e_n,e_n\rangle}{(a_n+2b_n)^2 b_n\langle e_n,e_n\rangle} = \frac{4(a_n+b_n)b_n}{(a_n+2b_n)^2}\cdot\frac{\langle A^2e_n,e_n\rangle^3}{\langle A^3e_n,e_n\rangle^2\langle e_n,e_n\rangle}$$

$$\ge \frac{256m^3M^3}{(M+m)^6}\cdot\frac{(a_n+b_n)b_n}{(a_n+2b_n)^2} . \tag{30}$$

Because $a_n \le M$ and $b_n \ge m$, we have $\frac{a_n}{b_n} \le \frac{M}{m}$. For $t \ge 0$ the function $f(t) = \frac{t+1}{(t+2)^2}$ is monotonically decreasing. Thus

$$\frac{(a_n+b_n)b_n}{(a_n+2b_n)^2} = \frac{\frac{a_n}{b_n}+1}{\left(\frac{a_n}{b_n}+2\right)^2} \ge \frac{\frac{M}{m}+1}{\left(\frac{M}{m}+2\right)^2} = \frac{(M+m)m}{(M+2m)^2} .$$

Using this, inequality (30) now becomes

$$\frac{4(a_n+b_n)\langle A^2e_n,e_n\rangle}{(a_n+2b_n)^2 b_n\langle e_n,e_n\rangle} \ge \frac{256m^4M^3}{(M+m)^5(M+2m)^2} .$$

Thus (29) shows the convergence of $\{x_n\}$ to x^* at the rate claimed. ∎

The statement of the theorem for the case $c=1$ is identical to Theorem (27), except that the ratio for the speed of convergence is given by the expression

$$\left(1 - \frac{64(m+2M)m^4M^3}{(M+m)^8}\right)^{1/2} .$$

The techniques of the proof are the same.

To conclude this section, we discuss a generalization of the method of steepest descent for unbounded Kpd operators. This method is due to Petryshyn [116], and its proof is based on an application of the basic method after suitable renorming in the sense of Petryshyn's generalization of the techniques of Friedrichs. We wish to solve (1) where A is a densely defined unbounded Kpd operator. In order to do this, we shall select an auxiliary Kpd operator L such that $D(A) \subset D(L)$ and for which the equation $Lx=y$, $y \in \mathcal{H}$, is easily solvable, at least for certain $y \in \mathcal{H}$. Also we choose L so that for some constants $M > m > 0$ we have for all $u \in D(A)$:

$$m\langle Lu,Ku\rangle \leq \langle Au,Ku\rangle \leq M\langle Lu,Ku\rangle . \tag{31}$$

We shall make use of the space $F(A,K)$ with the usual inner product

$$\langle u,v\rangle_K = \langle Au,Kv\rangle ,$$

and we assume that the usual extensions of A and K to all of $F(A,K)$ have been carried out. In addition we shall use the space $F(L,K)$. The inner product for this space will be denoted by

$$\langle u,v\rangle_L = \langle Lu,Kv\rangle ,$$

and we assume that L has also been extended. The inequalities in (31) show that the two associated norms $\|\cdot\|_K$ and $\|\cdot\|_L$ are equivalent.

By (31) and the Riesz Representation Theorem, there exists a bounded linear operator B on $F(L,K)$ such that

$$\langle u,v\rangle_K = \langle Bu,v\rangle_L \,.$$

If $Bu=0$, then $\langle u,v\rangle_K=0$ for all $v \in F(A,K) \subset F(L,K)$. Thus $u=0$ and B^{-1} exists. In fact, inequality (31) shows that

$$m\langle u,u\rangle_L \leq \langle Bu,u\rangle_L \leq M\langle u,u\rangle_L \,.$$

Thus B is bounded and positive bounded below on $F(L,K)$, and B^{-1} is bounded on $F(L,K)$.

Consider now the equation

$$Lu = Aw \,, \tag{33}$$

where $w \in D(A)$ is given. This is equivalent to the equation $\langle u,v\rangle_L = \langle w,v\rangle_K$ for all $v \in F(L,K)$. This equation is defined for all $u, v \in F(L,K)$. If for a given $w \in D(A)$, u^* solves this equation for all $v \in F(L,K)$, then u^* is a generalized solution of (33). Equation (32) shows that $\langle u^*,v\rangle_L = \langle Bw,v\rangle_L$ for all $v \in F(L,K)$. Thus $u^*=Bw$.

Now the equation

$$Lv = Aw - y \,, \quad y \in \mathcal{N} \,, \tag{34}$$

with $y_1 = L^{-1}y$, can be written $L(v+y_1)=Aw$ which is of the form (33). Thus (34) has a solution $v^*=Bw-y_1$. Equation (1) can be written in the equivalent form

$$\langle Bx,v\rangle_L = \langle y_1,v\rangle_L , \quad v \in F(L,K) ,$$

which is itself equivalent to the equation

$$Bx = y_1 , \quad y_1 \in F(L,K) . \tag{35}$$

Thus we shall be able to apply the method of steepest descent to this equation in $F(L,K)$. We shall refer to this method as the generalized method of steepest descent.

Theorem. (36)

Let A be a densely defined, unbounded Kpd operator in $\mathcal{H}$. If there exist a Kpd operator L on $D(A)$ and constants $M > m > 0$ such that (31) holds, then the generalized method of steepest descent produces a sequence $\{x_n\}$ which converges in $F(L,K)$ and in $\mathcal{H}$ to the unique (possibly generalized) solution x^* of (1) at least as fast as a geometric progression with ratio

$$\frac{M-m}{M+m} .$$

Proof.

Let $x_0 \in D(A)$ be an arbitrary initial approximation for the solution of (1). Let $z_0=y_1-Bx_0$ be the residual with respect to equation (35). Then $x_1=x_0+\alpha_0 z_0$, where

$$\alpha_0 = \frac{\langle z_0,z_0\rangle_L}{\langle Bz_0,z_0\rangle_L} = \frac{\langle z_0,z_0\rangle_L}{\langle z_0,z_0\rangle_K} \tag{37}$$

by (32). Note that for $r_0=y-Ax_0$ we have $Lz_0=r_0$. From a practical standpoint, the operator B does not appear. But we have seen that we can solve $Lz_0=r_0$ for z_0. Then we obtain α_0 from the second part of (37) and, finally, we can compute x_1. In general we let

$$x_{n+1} = x_n + \alpha_n z_n ,$$

where z_n is the generalized solution of $Lz=r_n$ in $F(L,K)$, and

$$\alpha_n = \frac{\langle z_n, z_n\rangle_L}{\langle z_n, z_n\rangle_K} .$$

The convergence in $F(L,K)$ now follows from Theorem (2). Since $\|u\|_L^2 = \langle Lu, Ku\rangle \geq \gamma\|u\|^2$, we also have convergence in $\mathcal{H}$. The estimate for the rate of convergence follows from Theorem (15) and its corollary. ▮

A recent method due to Peklova [115] applies gradient techniques to Sokolov's method [142, 143] for solving the linear operator equation of the second kind, $(I-A)x=y$. Peklova's result was mentioned in Section IV.1 in conjunction with successive approximation type methods. The reader is now in a position to note that Peklova's method combines successive approximation techniques and steepest descent techniques.

We also wish to mention a most interesting technique due to Vorobev [43, 44] although it really lies outside the scope of this paper. Vorobev uses the steepest descent iteration given by (3). Formula (4) shows that for a bounded and positive bounded below operator A , $0 < mI \leq A \leq MI$, we always have $\alpha_n \in [m,M]$ in the standard method of steepest descent. Most steepest descent methods differ from the standard method only in their choice of the α_n . Vorobev chooses the α_n from the interval $[m,M]$ at random. Thus he applies Monte Carlo techniques to the method of steepest descent. In another paper [45], Vorobev uses his probabilistic techniques on the alternating direction method of Kellogg, which we discussed in Section III.4. We do not discuss Vorobev's results. However, we do comment that of all the techniques which have been considered for use in solving operator equations, Monte Carlo techniques seem to have been considered the least. The reader interested in this subject may consult the article by Curtiss [7] and his bibliography.

SECTION 3

CONJUGATE DIRECTION METHODS

Although it is relatively easy to apply, the method of steepest descent often displays a disappointingly slow rate of convergence. The conjugate gradient method was developed in order to obtain better rates of convergence. In this section we shall discuss the conjugate direction methods with particular emphasis on the conjugate gradient method which, as we shall demonstrate, is a special conjugate direction method. It will become clear that the general conjugate direction method is largely of theoretical interest except in finite-dimensional spaces. However, the special case of the conjugate gradient method will be of practical interest as well.

Our proof of the convergence of the basic method and the estimate for the rate of convergence will follow Hayes [71] with certain modifications due to Daniel [60]. It turns out that both the conjugate gradient method and the method of steepest descent are special cases of the result of Samanskii given in Section II.3. Using a combination of the results of Samanskii [135] and Daniel [60], this fact will allow us to obtain an improved estimate for the rate of convergence of the conjugate gradient method. Also it will allow us to compare the method of steepest descent with the conjugate gradient method. We shall be able to conclude that, in general, the method of conjugate gradients seems to give faster convergence. The section will conclude with a consideration of Altman's work [50] on the method of conjugate gradients for operator equations in which the operator is unbounded.

Let A be a bounded and positive bounded below linear operator on a complex Hilbert space $\mathcal{H}$. We shall first seek a solution of the equation

$$Ax = y\ , \quad y \in \mathcal{H}\ , \tag{1}$$

by the conjugate direction method. In this method we again use the iterative scheme (1.12). The sequence $\{p_n\}$ of nonzero elements is chosen subject only to the condition

$$\langle p_i, Ap_j\rangle = 0\ , \quad i \neq j\ . \tag{2}$$

In a sense the method is not precise, for a great deal of latitude exists in choosing the sequence $\{p_n\}$. We note, however, that condition (2) is sufficient to insure that the vectors p_n are linearly independent. Before we can proceed, we require a lemma which gives some simple properties of the vectors p_n and r_n.

Lemma. (3)

The vectors p_n and r_n of the conjugate direction method satisfy:

(a) $r_i = r_{i-1} - \alpha_{i-1} Ap_{i-1}\ , \quad i=1,2,3,\dots\ .$

(b) $\langle r_j, p_i\rangle = 0\ , \quad i < j\ .$

(c) $\langle r_i, p_i \rangle = \langle r_{i-1}, p_i \rangle = \dots = \langle r_0, p_i \rangle$.

(d) $\alpha_i = \frac{\langle r_0, p_i \rangle}{\langle p_i, Ap_i \rangle}$.

Proof.

(a) This follows at once from (1.12).

(b) The proof is by induction on j with i fixed. For $j=i+1$ we have, using (a):

$$\langle r_{i+1}, p_i \rangle = \langle r_i - \alpha_i A p_i, p_i \rangle$$
$$= \langle r_i, p_i \rangle - \alpha_i \langle Ap_i, p_i \rangle = 0 .$$

If we assume (b) for $j=k-1 \geq i+1$, then

$$\langle r_k, p_i \rangle = \langle r_{k-1} - \alpha_{k-1} A p_{k-1}, p_i \rangle$$
$$= \langle r_{k-1}, p_i \rangle - \alpha_{k-1} \langle Ap_{k-1}, p_i \rangle$$
$$= \langle r_{k-1}, p_i \rangle = 0 .$$

(c) We note that

$$\langle r_i, p_i \rangle = \langle r_{i-1} - \alpha_{i-1} A p_{i-1}, p_i \rangle = \langle r_{i-1}, p_i \rangle ,$$

so that the result follows by induction.

(d) This follows from (c) and the definition of α_i . ▌

We are now ready to give the proof of convergence for the conjugate direction method. Although the proof follows Hayes [71], we modify the statement of the theorem somewhat, just as Daniel did in his dissertation [60].

Theorem. (4)

Let A be a bounded and positive bounded below operator. Let $\{p_n\}$ satisfy (2). If $B_n = sp[p_0, p_1, p_2, \ldots, p_{n-1}]$ and $B = \overline{\bigcup_{n=1}^{\infty} B_n}$, then the sequence $\{x_n\}$ of the conjugate direction method converges to a point $x' \in B$ such that $\langle r', z \rangle = 0$ for all $z \in B$. If $B = \mathcal{X}$, then $x' = x^*$.

Proof.

Although the choice $x_0 = 0$ is often undesirable in an actual problem, we can assume without loss of generality that $x_0 = 0$. Solving (1) with $x_0 \neq 0$ is equivalent to solving $Ax = y - Ax_0$ with zero as the initial approximation. Let $z \in B_n$ be arbitrary. Then $z = \beta_0 p_0 + \beta_1 p_1 + \ldots + \beta_{n-1} p_{n-1}$ for some $\beta_i \in \mathbb{C}$ and $x_n = \alpha_0 p_0 + \alpha_1 p_1 + \ldots + \alpha_{n-1} p_{n-1}$. Since

$$\langle r_n, z - x_n \rangle = \sum_{i=0}^{n-1} \overline{(\beta_i - \alpha_i)} \langle r_n, p_i \rangle = 0$$

by (b) of Lemma (3), we have for $z \neq x_n$:

$$\begin{aligned} E(z) - E(x_n) &= E(x_n + (z - x_n)) - E(x_n) \\ &= \langle z - x_n, A(z - x_n) \rangle - \langle r_n, z - x_n \rangle - \langle z - x_n, r_n \rangle \\ &= \langle z - x_n, A(z - x_n) \rangle > 0 . \end{aligned}$$

Thus x_n minimizes $E(x)$ on B_n. The result now follows from Lemma (1.10). ∎

Antoseiwicz and Rheinboldt [52] have pointed out that conditions are not known which would enable one to select x_0 in such a way that one could be sure that $B=\mathcal{H}$. In a practical problem only the first n elements of the sequence $\{p_i\}$ are actually computed. Therefore if $B=\mathcal{H}$ were made a hypothesis for Theorem (4), the theorem would be difficult to apply except in finite-dimensional spaces. Daniel [60] gives the following example to show that if $B \neq \mathcal{H}$, then it may well be that $x' \neq x^*$. Suppose that $sp[p_0', p_1', p_2', \ldots] = \mathcal{H}$. Discard p_0' and use the sequence $\{p_i\}$ with $p_i = p_{i+1}'$. Select $x_0 = x^* - p_0'$. Then for any $n \geq 0$:

$$\alpha_n = \frac{\langle r_n, p_n\rangle}{\langle p_n, Ap_n\rangle} = \frac{\langle y - A(x_0 + \sum_{i=0}^{n-1} \alpha_i p_i), p_n\rangle}{\langle p_n, Ap_n\rangle}$$

$$= \frac{\langle y - Ax^* + Ap_0' + \sum_{i=0}^{n-1} \alpha_i Ap_i, p_n\rangle}{\langle p_n, Ap_n\rangle}$$

$$= \frac{\langle Ap_0', p_n\rangle + \sum_{i=0}^{n-1} \alpha_i \langle Ap_i, p_n\rangle}{\langle p_n, Ap_n\rangle} = 0 .$$

Hence $x_0 = x_1 = x_2 = \ldots = x^* - p_0' \neq x^*$. These remarks provide justification for our earlier comment that the general conjugate direction method is mainly of theoretical interest in infinite-dimensional spaces.

As we shall see shortly, the conjugate gradient method is a special case of the conjugate direction method. Thus Theorem (4) will hold for that method as well. However, we shall find that a much more satisfying result holds for the conjugate gradient method which makes the method of practical interest. The idea for this result is due to Daniel [60]. We shall show that for the conjugate gradient method we always have $x_n \to x^*$, even when $B \neq \mathcal{N}$. This is one reason why we shall direct our attention to the conjugate gradient method for the rest of this section.

In the conjugate gradient method the p_n are chosen at each step. The procedure is:

$$x_{n+1} = x_n + \alpha_n p_n \ , \quad x_0 \text{ arbitrary}, \quad \alpha_n = \frac{\langle r_n, r_n \rangle}{\langle p_n, Ap_n \rangle} \ ,$$

$$p_{n+1} = r_{n+1} + \beta_n p_n, \ p_0 = r_0, \quad \beta_n = \frac{\langle r_{n+1}, r_{n+1} \rangle}{\langle r_n, r_n \rangle} \ . \tag{5}$$

It is not immediately obvious that condition (2) is satisfied, nor is it obvious that our definition of α_n coincides with the required one given in (1.12). To take care of this, and to prove other basic relations between the p_n and r_n , we have the following technical lemma due to Hestenes and Stiefel [24].

<u>Lemma.</u> (6)

The vectors p_n and r_n of the conjugate gradient method satisfy:

(a) $\langle p_i, Ap_j \rangle = 0$, $i \neq j$.

(b) $\langle r_i, r_j \rangle = 0$, $i \neq j$.

(c) $\langle r_j,p_i\rangle = \|r_i\|^2$, $i \geq j$.

(d) $\langle r_i,Ap_j\rangle = \langle p_i,Ap_j\rangle$, $i \neq j+1$.

(e) $$\alpha_i = \frac{\|r_i\|^2}{\langle p_i,Ap_i\rangle} = \frac{\langle p_i,r_i\rangle}{\langle p_i,Ap_i\rangle} = \frac{\langle p_i,r_0\rangle}{\langle p_i,Ap_i\rangle} .$$

(f) $$\beta_i = \frac{\|r_{i+1}\|^2}{\|r_i\|^2} = -\frac{\langle r_{i+1},Ap_i\rangle}{\langle p_i,Ap_i\rangle} = -\frac{\langle r_{i+1},Ar_i\rangle}{\langle p_i,Ap_i\rangle} , \quad r_i \neq 0 .$$

(g) $$\frac{\|r_i\|^2}{\langle r_i,Ar_i\rangle} < \alpha_i < \frac{\|p_i\|^2}{\langle p_i,Ap_i\rangle} , \quad r_i \neq 0 .$$

Proof.

(a)(b)(c) We shall prove (a), (b), and (c) simultaneously by induction. It will be sufficient to show that

(a)' $\langle p_n,Ap_j\rangle = 0$, $j < n$,

(b)' $\langle r_j,r_{n+1}\rangle = 0$, $j < n+1$,

(c)' $\langle r_j,p_n\rangle = \|r_n\|^2$, $j < n+1$,

hold for all n . For $n=0$, condition (a)' is satisfied vacuously. Since $\langle r_0,r_1\rangle = \langle r_0,r_0-\alpha_0 Ap_0\rangle = \langle r_0,r_0\rangle - \alpha_0\langle p_0,Ap_0\rangle = 0$, condition (b)' is satisfied. As for (c)', we need only note that $r_0=p_0$. Thus we now assume that all three of these statements hold for $n=k-1$. First of all, for any m ,

$$\begin{aligned} p_m &= r_m + \beta_{m-1}p_{m-1} = r_m + \beta_{m-1}(r_{m-1} + \beta_{m-2}p_{m-2}) \\ &= \dots = r_m + \beta_{m-1}r_{m-1} + \beta_{m-1}\beta_{m-2}r_{m-2} + \dots + \beta_{m-1} \dots \beta_0 r_0 \\ &= \frac{\langle r_m,r_m\rangle}{\langle r_m,r_m\rangle} r_m + \frac{\langle r_m,r_m\rangle}{\langle r_{m-1},r_{m-1}\rangle} r_{m-1} + \dots + \frac{\langle r_m,r_m\rangle}{\langle r_0,r_0\rangle} r_0 \end{aligned}$$

$$= \|r_m\|^2 \sum_{i=0}^{m} \frac{r_i}{\langle r_i, r_i \rangle} .$$

Using this formula together with the induction hypothesis:

$$\langle r_j, p_k \rangle = \|r_k\|^2 \sum_{i=0}^{k} \frac{\langle r_j, r_i \rangle}{\langle r_i, r_i \rangle} = \|r_k\|^2 , \quad j < k+1 ,$$

which shows that (c)' holds when $n=k$. Now, for $j \le k$,

$$\begin{aligned} \langle r_j, p_k \rangle &= \langle r_{j-1} - \alpha_{j-1} A p_{j-1}, p_k \rangle \\ &= \langle r_{j-1}, p_k \rangle - \alpha_{j-1} \langle A p_{j-1}, p_k \rangle . \end{aligned}$$

Using (c)' this may be written

$$\|r_k\|^2 = \|r_k\|^2 - \alpha_{j-1} \langle A p_{j-1}, p_k \rangle .$$

Unless $\alpha_{j-1}=0$ (in which case $r_{j-1}=0$ and x^* was obtained on the previous iteration), we have $\langle p_{j-1}, A p_k \rangle = 0$ with $j-1 < k$. This is equivalent to $\langle p_k, A p_j \rangle = 0$ for $j < k$ which shows that (a)' holds for $n=k$. Finally, if $j < k$, then using (d) $\langle r_j, r_{k+1} \rangle = \langle r_j, r_k \rangle - \alpha_k \langle r_j, A p_k \rangle = 0$. If $j=k$, we note that $\langle p_k, A p_k \rangle = \langle r_k + \beta_{k-1} p_{k-1}, A p_k \rangle = \langle r_k, A p_k \rangle - \beta_{k-1} \langle p_{k-1}, A p_k \rangle = \langle r_k, A p_k \rangle$ so that

$$\begin{aligned} \langle r_k, r_{k+1} \rangle &= \langle r_k, r_k \rangle - \alpha_k \langle r_k, A p_k \rangle \\ &= \langle r_k, r_k \rangle - \alpha_k \langle p_k, A p_k \rangle = 0 . \end{aligned}$$

Thus (b)' also holds for $n=k$, and the induction is completed.

(d) For $i \neq j+1$:

$$\langle p_i, Ap_j\rangle = \langle r_i + \beta_{i-1}p_{i-1}, Ap_j\rangle$$

$$= \langle r_i, Ap_j\rangle + \beta_{i-1}\langle p_{i-1}, Ap_j\rangle = \langle r_i, Ap_j\rangle .$$

(e) We have defined α_i by the first expression. The middle expression follows from (c). At this point we note that this shows our original expression (1.12) for α_n actually holds. Since we have shown in (a) that the conjugate gradient method is a conjugate direction method, we can now use all the properties of Lemma (3) in conjunction with the conjugate gradient method. Thus the last expression for α_i follows from (d) of Lemma (3).

(f) We have defined β_i by the first expression. Note that α_i and β_i are real by definition, and that $\langle r_i, Ap_j\rangle$ is real for $i \neq j+1$ by (d). Thus

$$\frac{\|r_{i+1}\|^2}{\|r_i\|^2} = \frac{\langle r_{i+1}, p_{i+1}\rangle}{\langle r_i, p_i\rangle} = \frac{\langle r_{i+1}, r_{i+1}+\beta_i p_i\rangle}{\langle r_i, p_i\rangle}$$

$$= \frac{\langle r_{i+1}, r_i - \alpha_i Ap_i + \beta_i p_i\rangle}{\langle r_i, p_i\rangle}$$

$$= \frac{-\alpha_i\langle r_{i+1}, Ap_i\rangle}{\langle r_i, p_i\rangle} = -\frac{\langle r_{i+1}, Ap_i\rangle}{\langle p_i, Ap_i\rangle} .$$

This shows that the middle expression for β_i is correct. Finally, $\langle r_{i+1}, Ap_i\rangle = \langle Ar_{i+1}, p_i\rangle = \langle Ar_{i+1}, r_i + \beta_{i-1}p_{i-1}\rangle = \langle r_{i+1}, Ar_i\rangle + \beta_{i-1}\langle r_{i+1}, Ap_{i-1}\rangle = \langle r_{i+1}, Ar_i\rangle$. Therefore the final expression for β_i is valid.

(g) First of all, using (b) of Lemma (3):

$$\|p_i\|^2 = \|r_i+\beta_{i-1}p_{i-1}\|^2 = \|r_i\|^2 + 2\beta_{i-1}\langle r_i,p_{i-1}\rangle+\beta_{i-1}^2\|p_{i-1}\|^2$$
$$= \|r_i\|^2 + \beta_{i-1}^2\|p_{i-1}\|^2 > \|r_i\|^2 .$$

If $\beta_{i-1}=0$, then (f) shows $r_i=0$. Thus

$$\frac{\|p_i\|^2}{\langle p_i,Ap_i\rangle} = \frac{\|r_i\|^2}{\langle p_i,Ap_i\rangle} \cdot \frac{\|p_i\|^2}{\|r_i\|^2} > \frac{\|r_i\|^2}{\langle p_i,Ap_i\rangle} = \alpha_i .$$

Now,

$$\langle r_i,Ar_i\rangle = \langle p_i-\beta_{i-1}p_{i-1},A(p_i-\beta_{i-1}p_{i-1})\rangle$$
$$= \langle p_i,Ap_i\rangle + \beta_{i-1}^2\langle p_{i-1},Ap_{i-1}\rangle > \langle p_i,Ap_i\rangle ,$$

so that

$$\frac{\|r_i\|^2}{\langle r_i,Ar_i\rangle} = \frac{\|r_i\|^2}{\langle p_i,Ap_i\rangle} \cdot \frac{\langle p_i,Ap_i\rangle}{\langle r_i,Ar_i\rangle} < \frac{\|r_i\|^2}{\langle p_i,Ap_i\rangle} = \alpha_i . \blacksquare$$

<u>Theorem.</u> (7)

Let A be a bounded and positive bounded below operator, $0 < mI \leq A \leq MI$. The sequence $\{x_n\}$ of the conjugate gradient method converges to x^* at least as fast as a geometric progression with ratio

$$\frac{M - m}{M + m} .$$

<u>Proof.</u>

As usual we let $B_n = sp[p_0,p_1,\ldots,p_{n-1}]$. The vector x_n still minimizes $E(x)$ on B_n since this is a conjugate direction method. Note, however, that we do not require that $B=\mathcal{H}$. We know that $\{E(x_n)\}$ converges, hence is Cauchy. Thus by (1.11) and (c) of Lemma (6),

$$E(x_n) - E(x_{n+1}) = E(x_n) - E(x_n+\alpha_n p_n)$$

$$= \frac{\langle r_n,p_n\rangle^2}{\langle p_n,Ap_n\rangle} = \alpha_n\|r_n\|^2 \to 0 . \tag{8}$$

But (g) of Lemma (6) shows that $\alpha_n \geq \frac{1}{M} > 0$. Hence we must conclude that $r_n \to 0$ or $x_n \to x^*$.

Using $E(x_n)-E(x_{n+1})=\alpha_n\|r_n\|^2$ from (8) and $E(x_n)= \frac{\langle r_n,A^{-1}r_n\rangle}{\langle r_n,r_n\rangle}\|r_n\|^2$,

we obtain

$$E(x_{n+1}) = E(x_n) - E(x_n)\left(\frac{\alpha_n\langle r_n,r_n\rangle}{\langle r_n,A^{-1}r_n\rangle}\right) .$$

But $\alpha_n > \frac{\langle r_n,r_n\rangle}{\langle r_n,Ar_n\rangle}$ by (g) of Lemma (6). Thus

$$E(x_{n+1}) < E(x_n)\left(1 - \frac{\langle r_n,r_n\rangle^2}{\langle r_n,A^{-1}r_n\rangle\langle r_n,Ar_n\rangle}\right)$$

$$\leq E(x_n)\left(1 - \frac{4mM}{(M+m)^2}\right)$$

$$= E(x_n)\left(\frac{M-m}{M+m}\right)^2 ,$$

by Theorem (II.2.7). Since $E(x_n) \geq m\|x_n-x^*\|^2$, we obtain the desired ratio. ▌

At this point we shall establish the connection between the result of Samanskii (Theorem (II.3.13)) and the conjugate gradient method. Although the proof of Samanskii's result is valid in a complex Hilbert space, the comparisons which we are about to make should be understood in the context of a real Hilbert space because of the use made of Chebyshev polynomials. These results are due to Daniel [60] and Samanskii [135].

Theorem. (9)

Let A be a bounded, self-adjoint, and positive bounded below operator on a real Hilbert space. Then the conjugate gradient method belongs to the class of algorithms of Samanskii of the form

$$x_{n+1} = x_0 - P_n(A)r_0 , \tag{10}$$

where $P_n(\lambda)$ is a polynomial of degree at most n.

Proof.

We proceed by induction. For $n=0$, $x_1=x_0+\alpha_0 p_0=x_0+\alpha_0 r_0$. Thus we can choose $P_0(\lambda) = -\alpha_0$. Assume that $x_n=x_0-P_{n-1}(A)r_0$. Then:

$$\begin{aligned}
x_{n+1} = x_n+\alpha_n p_n &= x_0-P_{n-1}(A)r_0 + \alpha_n(r_n+\beta_{n-1}p_{n-1}) \\
&= x_0-P_{n-1}(A)r_0 + \alpha_n\{A(x^*-x_n)+\beta_{n-1}p_{n-1}\} \\
&= x_0-P_{n-1}(A)r_0 + \alpha_n\{A(x^*-x_0+P_{n-1}(A)r_0)+\beta_{n-1}p_{n-1}\} \\
&= x_0-\{P_{n-1}(A)-\alpha_n AP_{n-1}(A)-\alpha_n\}r_0 + \alpha_n\beta_{n-1}p_{n-1} \\
&= x_0-P_{n,1}(A)r_0 + \alpha_n\beta_{n-1}p_{n-1}
\end{aligned}$$

$$= x_0 - P_{n,1}(A)r_0 + \alpha_n \beta_{n-1}(r_{n-1} + \beta_{n-2}p_{n-2})$$

$$= x_0 - P_{n,1}(A)r_0 + \alpha_n \beta_{n-1}\{A(x^* - x_0 + P_{n-2}(A)r_0) + \beta_{n-2}p_{n-2}\}$$

$$= x_0 - P_{n,1}(A)r_0 + \alpha_n \beta_{n-1}\{AP_{n-2}(A) + 1\}r_0 + \alpha_n \beta_{n-1}\beta_{n-2}p_{n-2}$$

$$= x_0 - P_{n,2}(A)r_0 + \alpha_n \beta_{n-1}\beta_{n-2}p_{n-2}$$

$$= \ldots = x_0 - P_{n,n}(A)r_0 + \alpha_n \beta_{n-1}\beta_{n-2}\cdots\beta_0 p_0$$

$$= x_0 - \{P_{n,n}(A) - \alpha_n \beta_{n-1}\beta_{n-2}\cdots\beta_0\}r_0$$

$$= x_0 - P_n(A)r_0 . \blacksquare$$

Now we can show even more. Not only does the conjugate gradient method lie in the class of algorithms considered by Samanskii, but it plays a very special role in this class. In a certain sense, about to be explained, the conjugate gradient method is optimal in this class. The proof is immediate when one considers that all conjugate direction methods minimize the functional $E(x)$ on $x_0 + B_n$ at the nth iteration.

<u>Theorem.</u> (11)

In a real Hilbert space, $E(x_n)$ is least for the x_n produced by the conjugate gradient method when compared with $E(x_n)$ for x_n obtained from any other method of the form (10).

At this point Samanskii chooses to use the Chebyshev polynomials as the polynomials in his method. We do not go into detail. However, by using them, he is able to obtain the following estimate:

$$E(x_n) \leq \left(\frac{2(1-\alpha)^n}{(1+\sqrt{\alpha})^{2n} + (1-\sqrt{\alpha})^{2n}}\right)^2 E(x_0) \ , \tag{12}$$

where $\alpha = \frac{m}{M}$ and m and M are respectively the greatest lower and least upper bounds for $\sigma(A)$. Noting that $1-\alpha = (1+\sqrt{\alpha})(1-\sqrt{\alpha})$ and using

$$\frac{(ab)^n}{a^{2n} + b^{2n}} \leq \left(\frac{b}{a}\right)^n \ , \qquad 0 \leq b < a \ ,$$

with $a = 1 + \sqrt{\alpha}$ and $b = 1 - \sqrt{\alpha}$, we can write

$$E(x_n) \leq 4\left(\frac{1 - \sqrt{\alpha}}{1 + \sqrt{\alpha}}\right)^{2n} E(x_0) \ .$$

Theorems (9) and (11) allow us to conclude that the conjugate gradient method converges at least as fast as a geometric progression with ratio

$$\frac{1 - \sqrt{\alpha}}{1 + \sqrt{\alpha}} = \frac{\sqrt{M} - \sqrt{m}}{\sqrt{M} + \sqrt{m}} \ .$$

We note that $\frac{m}{M} \leq \sqrt{\frac{m}{M}}$. Hence $1 - \sqrt{\frac{m}{M}} \leq 1 - \frac{m}{M}$ and $1 + \sqrt{\frac{m}{M}} \geq 1 + \frac{m}{M}$. These two inequalities show that

$$\frac{M - m}{M + m} \geq \frac{\sqrt{M} - \sqrt{m}}{\sqrt{M} + \sqrt{m}} \ . \tag{13}$$

Strict inequality holds unless $m=M$, a case of little interest. Thus Samanskii's use of Chebyshev polynomials has allowed him to obtain an improved estimate for the rate of convergence of the method of conjugate gradients.

Next we show that the method of steepest descent also belongs to Samanskii's class of algorithms. This will allow us to compare the rate of convergence of the method of steepest descent and the conjugate gradient method.

Theorem. (14)

Let A be a bounded, self-adjoint, and positive bounded below operator on a real Hilbert space. Then the steepest descent method belongs to the class of algorithms of Samanskii of the form (10).

Proof.

Again, the proof is by induction. Since $x_1 = x_0 + \alpha_0 r_0$, the choice $P_0(\lambda) = -\alpha_0$ shows that the result holds for $n=0$. Assume that $x_n = x_0 - P_{n-1}(A) r_0$. Then:

$$
\begin{aligned}
x_{n+1} &= x_n + \alpha_n r_n = x_0 - P_{n-1}(A) r_0 + \alpha_n A(x^* - x_n) \\
&= x_0 - P_{n-1}(A) r_0 + \alpha_n A(x^* - x_0 + P_{n-1}(A) r_0) \\
&= x_0 - (P_{n-1}(A) - \alpha_n A P_{n-1}(A) - \alpha_n) r_0 \\
&= x_0 - P_n(A) r_0 . \blacksquare
\end{aligned}
$$

Having placed the method of steepest descent in this class of algorithms, we obtain the following corollary to Theorem (11).

<u>Corollary.</u> (15)

Let the operator A be as above. For all n, $E(x_n)$ for the x_n obtained from the conjugate gradient method is less than or equal to $E(x_n)$ for the x_n obtained by the method of steepest descent.

In spite of all that we have shown, because our estimates $\frac{M-m}{M+m}$ and $\frac{\sqrt{M}-\sqrt{m}}{\sqrt{M}+\sqrt{m}}$ are only upper bounds, we cannot conclude that the conjugate gradient method converges faster than the method of steepest descent. In view of the inequality (13), it does seem likely that, in general, the conjugate gradient method will give a faster rate of convergence than the method of steepest descent. We support this intuitive statement with two more facts.

If q_n is the ratio in (12) and q_n' is the ratio $\left(\frac{1-\alpha}{1+\alpha}\right)^n$, $\alpha = \frac{m}{M}$, associated with the method of steepest descent, then the ratio

$$\frac{q_n}{q_n'} = \frac{2(1-\alpha)^n}{(1+\sqrt{\alpha})^{2n} + (1-\sqrt{\alpha})^{2n}} \cdot \frac{(1+\alpha)^n}{(1-\alpha)^n}$$

$$\leq 2\left(\frac{1+\alpha}{(1+\sqrt{\alpha})^2}\right)^n = 2\left(\frac{1}{1+\frac{2\sqrt{\alpha}}{1+\alpha}}\right)^n$$

converges to zero at least as fast as a geometric progression.

Akaike [1] has given examples to show that one cannot improve the estimate $\frac{M-m}{M+m}$ for the method of steepest descent. That is, he gives examples in which the rate of convergence can be made to come arbitrarily close to this estimate. When $m < M$, we have noted that the inequality in (13) is strict. Thus, from his examples, we obtain examples in which we do have faster convergence for the conjugate gradient method.

The final topic that will be discussed in this section is the use of the conjugate gradient method for solving the equation (1) when the operator A is unbounded. First we point out that the techniques used in Theorem (2.36) by Petryshyn for the method of steepest descent can be applied to the conjugate gradient method. Doing this, we obtain a conjugate gradient method for unbounded Kpd operators in the form

$$x_{n+1} = x_n + \alpha_n p_n , \quad x_0 \text{ arbitrary},$$

$$\alpha_n = \frac{\langle z_n, z_n\rangle_L}{\langle p_n, Bp_n\rangle_L} = \frac{\langle z_n, z_n\rangle_L}{\langle z_n, z_n\rangle_K} ,$$

$$p_{n+1} = z_{n+1} + \beta_n p_n, \quad p_0 = z_0 , \quad \beta_n = \frac{\langle z_{n+1}, z_{n+1}\rangle_L}{\langle z_n, z_n\rangle_L} ,$$

where z_{n+1} is the generalized solution of $Lz=r_{n+1}$ in $F(L,K)$. Both Daniel [60] and Hayes [71] have indicated algorithms for unbounded operators. These schemes are also based on the renorming techniques of Friedrichs.

Here we shall give a technique due to Altman [50, p. 33]. It is of particular interest because it shows a connection between the Ritz method [108, pp. 85-95] and the conjugate gradient method. The standard iteration formula (1.12) remains unchanged in Altman's method. Let A be a self-adjoint and positive bounded below operator. One should note that the proofs of Lemmas (3) and (6) did not depend on the boundedness of the operator A. Thus one can use all of those properties of the p_i and r_i vectors even when A is unbounded. The formula $x_{n+1} = x_n + \alpha_n p_n$ may be written

$$x_{n+1} = x_0 + \sum_{i=0}^{n} \alpha_i p_i . \tag{16}$$

Using (c) of Lemma (3):

$$\langle r_i, p_i \rangle = \langle r_0, p_i \rangle = \langle A(x^* - x_0), p_i \rangle$$
$$= \langle x^* - x_0, Ap_i \rangle .$$

Thus we may write α_i in the form

$$\alpha_i = \frac{\langle r_i, p_i \rangle}{\langle p_i, Ap_i \rangle} = \frac{\langle x^* - x_0, Ap_i \rangle}{\langle p_i, Ap_i \rangle} .$$

Using the space F(A) of Friedrichs, we can now write α_i in the form

$$\alpha_i = \frac{\langle x^* - x_0, p_i \rangle_A}{\langle p_i, p_i \rangle_A} , \tag{17}$$

so that the substitution in (16) produces the formula

$$x_{n+1} = x_0 + \sum_{i=0}^{n} \frac{\langle x^* - x_0, p_i \rangle_A}{\langle p_i, p_i \rangle_A} . \tag{18}$$

If x^*-x_0 can be approximated in $F(A)$ with arbitrary accuracy by expressions of the form $\sum_{i=0}^{m} \gamma_i p_i$, then substitution of this expression for x^*-x_0 in (18) shows that $x_{n+1}-x_0 \cong x^*-x_0$. That is, $x_n \to x^*$ in $F(A)$.

We have seen in the proof of part (c) of Lemma (6) that p_n can be expressed as a linear combination of the r_i , $0 \leq i \leq n$. Thus (a) of Lemma (3) shows that p_n can be expressed as a linear combination of elements of the form $A^i r_0$, $0 \leq i \leq n$, provided these expressions are well-defined; i.e., provided $r_0 \in D(A)$, $Ar_0 \in D(A)$, $A^2 r_0 \in D(A)$, etc. Therefore, if x^*-x_0 can be approximated in $F(A)$ by linear combinations of elements of the form $A^i r_0$, $i=0,1,2,\ldots$, such expressions being well-defined, then $x_n \to x^*$ in $F(A)$.

<u>Theorem.</u> (19)

Let A be a densely defined and positive bounded below linear operator in $\mathcal{H}$. If r_0 can be approximated in $\mathcal{H}$ with arbitrary accuracy by meaningful linear combinations of elements of the form $A^i r_0$, $i=1,2,3,\ldots$, then the sequence $\{x_n\}$ generated by the conjugate gradient method converges to the unique solution x^* of (1).

<u>Proof.</u>

We have already observed in Theorem (I.1.7) that, for the type of operator A considered here, equation (1) has a unique solution x^* , possibly in the generalized sense. Let $\epsilon > 0$ be given. By hypothesis, for some n we can find $\gamma_1, \gamma_2, \gamma_3, \ldots, \gamma_n \in \mathbb{C}$ such that

$$\|r_0 - \sum_{i=1}^{n} \gamma_i A^i r_0\| < \epsilon .$$

Since A is positive bounded below, $\|Au\| \geq m\|u\|$ for some $m > 0$. Thus

$$\|(x^*-x_0) - \sum_{i=1}^{n} \gamma_i A^{i-1} r_0\| < \frac{\epsilon}{m} .$$

But $\|u\|_A^2 = \langle Au,u\rangle \leq \|Au\| \, \|u\|$. Multiplication of the two inequalities above yields

$$\|(x^*-x_0) - \sum_{i=1}^{n} \gamma_i A^{i-1} r_0\|_A^2 < \frac{\epsilon^2}{m} .$$

The comment immediately preceeding the statement of the theorem shows that $x_n \to x^*$ in $F(A)$. The inequality (I.1.5) shows that $x_n \to x^*$ in $\mathcal{H}$ also. ▌

In the space $F(A)$, the collection $\{p_i\}$ is an orthogonal collection. Equation (18) shows that the vector $x_{n+1}-x_0$ which we have constructed coincides with a Ritz approximate solution for x^*-x_0, using the collection $\{p_i\}$ as coordinates. Notice that $x_{n+1}-x_0 = P_{n+1}x^*-x_0$ where P_{n+1} is the orthogonal projection of $F(A)$ onto the subspace $B_{n+1} = sp[p_0, p_1, \ldots, p_n]$. However, we obtain the vector p_n only at the $n^{\underline{th}}$ step here, and we cannot use (17) to compute the $n^{\underline{th}}$ Ritz coefficient without actually knowing $x^* - x_0$. Thus they coincide only on a theoretical basis.

We close this section by mentioning another article by Kammerer and Nashed [74]. In this article the steepest descent results of Nashed [110] and of Kammerer and Nashed [72] mentioned in the introduction to Section 2 are extended to the conjugate gradient method. Since their article [74] also deals with equations (1) which are not solvable, we shall not discuss these results either.

BIBLIOGRAPHY

The first section of this bibliography contains general, historical, and background references. It is not intended to be complete in any way. The second section contains the current literature for the topics which are covered in this work. It is hoped that this section is fairly complete. Inevitably, there will occur some inadvertent omissions. In addition, the question of what is "current" is a matter of opinion. Whenever available, the applicable review from the American Mathematical Society's Mathematical Reviews is listed.

SECTION 1

1. H. Akaike, On a successive transformation of probability distribution and its application to the analysis of the optimum gradient method, Ann. Inst. Statist. Math. Tokyo 11 (1959), 1-16. MR 21 #6694.

2. G. Birkhoff, The mean ergodic theorem, Duke Math. J. 5 (1939), 19-20.

3. F. Browder, On the iteration of transformations in non-compact minimal dynamical systems, Proc. Amer. Math. Soc. 9 (1958), 773-780. MR 20 #3456.

4. A. Cauchy, Méthode générale pour la résolution des systèmes d'équations simultanées, C. R. Acad. Sci. Paris 25 (1847) 536-538.

5. R. Courant, Variational methods for the solution of problems of equilibrium and vibrations, Bull. Amer. Math. Soc. 49 (1943), 1-23. MR 4, 200.

6. R. Courant and D. Hilbert, Methods of Mathematical Physics, 2 vols., Interscience, New York, 1953 and 1962. MR 16, 426 (vol. I). MR 25 #4216 (vol. II).

7. J. Curtiss, "Monte Carlo" methods for the iteration of linear operators, J. Math. and Phys. 32 (1953), 209-232. MR 15, 560.

8. N. Dunford and J. Schwartz, Linear Operators, Parts I, II, and III, Pure and Applied Mathematics, vol. 7, Interscience, New York, 1958, 1963, 1971. MR 22 #8302, MR 32 #6181.

9. W. Eberlein, Abstract ergodic theorems and weak almost periodic functions, Trans. Amer. Math. Soc. 67 (1949), 217-240. MR 12, 112.

10. M. Engeli, T. Ginsburg, H. Rutishauser, and E. Stiefel, Refined iterative methods for computation of the solution and the eigenvalues of self-adjoint boundary value problems, Mitt. Inst. Angew. Math. Zürich, no. 8, 1959. MR 26 #3218.

11. D. Faddeev and V. Faddeeva, Computational Methods of Linear Algebra, translated by R. Williams, W. H. Freeman, San Francisco, 1963. MR 28 #1742.

12. W. Feller, An Introduction to Probability Theory and Its Applications, vol. 1, 3rd ed., Wiley, New York, 1968. MR 37 #3604.

13. G. Forsythe, Solving linear algebraic equations can be interesting, Bull. Amer. Math. Soc. 59 (1953), 299-329. MR 15, 65.

14. L. Fox, H. Huskey, and J. Wilkinson, Notes on the solution of algebraic linear simultaneous equations, Quart. J. Mech. Appl. Math. 1 (1948), 149-173. MR 10, 152.

15. V. Fridman, Method of successive approximations for a Fredholm integral equation of the first kind, Uspehi Mat. Nauk 11 (1956), no. 1, 233-234. (Russian) MR 17, 861.

16. K. Friedrichs, Spektraltheorie halbbeschränkter Operatoren, Math. Ann. 109 (1934), 465-487.

17. W. Greub and W. Rheinboldt, On a generalization of an inequality of L. V. Kantorovich, Proc. Amer. Math. Soc. 10 (1959), 407-415. MR 21 #3774.

18. J. Hadamard, Mémoire sur le problème d'analyse relatif à l'équilibre des plaques élastiques encastrées, Mémoires présentées par divers savants étrangers à l'Académie des Sciences de l'Institut National de France, vol. 33, no. 4, 1907.

19. P. Halmos, A Hilbert Space Problem Book, D. Van Nostrand Co., Princeton, 1967. MR 34 #8178.

20. E. Hellinger and O. Toeplitz, Integralgleichungen und Gleichungen mit unendlichen Unbekannten, Encyklopädie der mathematischen Wissenschaften, II C 13, 1335-1597, B. G. Teubner, Leipzig, 1927.

21. M. Hestenes, The solution of linear equations by minimization, National Bureau of Standards NAML Report #52-45, 1951.

22. ________, Iterative computational methods, Comm. Pure Appl. Math. 8 (1955), 85-95. MR 16, 863.

23. ________, The conjugate-gradient method for solving linear systems, Proc. Symposia in Appl. Math. 6 (1956), 83-102. MR 18, 824.

24. M. Hestenes and E. Stiefel, Methods of conjugate gradients for solving linear systems, J. Res. Nat. Bur. Standards 49 (1952), 409-436. MR 15, 651.

25. L. Kantorovic, The method of successive approximations for functional equations, Acta Math. 71 (1939), 63-97. MR 1, 18.

26. M. Krasnoselskii, Topological Methods in the Theory of Nonlinear Integral Equations, Pergamon Press, Macmillan, New York, 1964. MR 20 #3464. MR 28 #2414.

27. M. Krasnoselskii and S. Krein, Iteration process with minimum residua, Mat. Sb. 31 (1952), 315-334. (Russian) MR 14, 692.

28. J. Liouville, Sur le dévelopment des fonctions ou parties des fonctions en séries dont les divers termes sont assujettis à satisfaire à une même équation différentielle du second ordre contenant un paramètre variable, J. Math. Pures Appl. 2 (1837), 16-35.

29. S. Mihlin, Sur un théorème de F. Noether, Dokl. Akad. Nauk SSSR (International Edition) 43 (1944), 139-141. MR 6, 157.

30. C. Müller, A new method for solving Fredholm integral equations, Comm. Pure Appl. Math. 8 (1955), 635-640. MR 17, 1215.

31. A. Naisul, Improvement of the convergence of successive approximation methods for linear equations, Soviet Math. Dokl. 158 (1964), 1229-1230. MR 29 #4190.

32. I. Natanson, On the theory of approximate solutions of equations, Leningrad. Gos. Ped. Inst. Ucen. Zap. 64 (1948). (Russian)

33. C. Neumann, Untersuchungen über das logarithmische und Newtonische Potential, Teubner, Leipzig, 1877.

34. D. Peaceman and H. Rachford, Jr., The numerical solution of parabolic and elliptic differential equations, SIAM J. Appl. Math. 3 (1955), 28-41. MR 17, 196.

35. Y. Pearlman, B. G. Galerkin's method in variational calculus and in the theory of elasticity, Prikl. Mat. Meh. 5 (1941), 345-358. (Russian) MR 4, 203.

36. G. Polya and G. Szegö, Aufgaben und Lehrsätze aus der Analysis, Erster Band, Springer, Berlin, 1925. MR 15, 512.

37. W. Reid, Symmetrizable completely continuous linear transformations in Hilbert space, Duke Math. J. 18 (1951), 41-56. MR 13, 564.

38. F. Riesz and B. Sz.-Nagy, Functional Analysis, translated by Leo F. Boron, Frederick Ungar, New York, 1955. MR 17, 175.

39. E. Stiefel, Über einige Methoden der Relaxationsrechnung, Z. Angew. Math. Phys. 3 (1952), 1-33. MR 13, 874.

40. F. Stummel, Rand- und Eigenwertaufgaben in Sobolewschen Räumen, Lecture Notes in Mathematics, vol. 102, Springer, Berlin, 1969.

41. A. Taylor, Introduction to Functional Analysis, Wiley, New York, 1958. MR 20 #5411.

42. G. Temple, The general theory of relaxation methods applied to linear systems, Proc. Roy. Soc. Ser. A 169 (1939), 476-500.

43. Ju. Vorobev, A random iterational process, USSR Comput. Math. and Math. Phys. 4 (1964), no. 6, 150-158. MR 30 #5462.

44. ________, A random iteration process II, USSR Comput. Math. and Math. Phys. 5 (1965), no. 5, 1-13. MR 34 #7004.

45. ________, A random iteration process in the method of alternating directions, USSR Comput. Math. and Math. Phys. 8 (1968), no. 3, 226-238. MR 38 #2960.

46. G. Wiarda, Integralgleichungen, Teubner, Leipzig and Berlin, 1930.

47. K. Yosida and S. Kakutani, Operator-theoretical treatment of Markoff's process and mean ergodic theorem, Ann. of Math. 42 (1941), 188-228. MR 2, 230.

SECTION 2

48. M. Altman, An approximate method for solving linear equations in Hilbert space, Bull. Acad. Polon. Sci. Ser. Sci. Math. Astronom. Phys. 5 (1957), 601-604. MR 19, 984.

49. ________, On the approximate solutions of operator equations in Hilbert space, Bull. Acad. Polon. Sci. Ser. Sci. Math. Astronom. Phys. 5 (1957), 605-609. MR 19, 984.

50. ________, Approximation Methods in Functional Analysis, Lecture Notes, Math. 107c, California Inst. Tech., Pasadena, 1959.

51. ________, A linear iterative method with a vector parameter, Bull. Acad. Polon. Sci. Ser. Sci. Math. Astronom. Phys. 9 (1961), 169-174. MR 24 #A2262a.

52. H. Antosiewicz and W. Rheinboldt, Conjugate-direction methods and the method of steepest descent, Survey of Numerical Analysis, J. Todd, ed., McGraw-Hill, New York, 1962, 501-512. MR 25 #1632.

53. G. Bessmertnyh, Two methods for approximate solution of operator equations in Hilbert space, Soviet Math. Dokl. 1 (1960), 211-214. MR 25 #431.

54. H. Bialy, Iterative Behandlung linearer Funktionalgleichungen, Arch. Rational Mech. Anal. 4 (1959), 166-176. MR 22 #1095.

55. M. Birman, Some estimates for the method of steepest descent, Uspehi Mat. Nauk 5 (1950), no. 3, 152-155. (Russian) MR 12, 32.

56. F. Browder and W. Petryshyn, The solution by iteration of linear functional equations in Banach spaces, Bull. Amer. Math. Soc. 72 (1966), 566-570. MR 32 #8155a.

57. A. Buledza, Acceleration of the convergence of iterative processes in an approximate solution of linear operator equations, Dopovidi Akad. Nauk Ukrain, RSR 1961, 265-269. (Russian) MR 23 #A523.

58. V. Carusnikov, Universally optimal iteration processes, Soviet Math. Dokl. 11 (1970), 1431-1435. MR 44 #4541.

59. ________, On optimal approximate methods for the solution of linear problems, Soviet Math. Dokl. 12 (1971), 1039-1043. MR 44 #2323.

60. J. Daniel, The conjugate gradient method for linear and nonlinear operator equations, Doctoral Thesis, Stanford University, Stanford, 1965.

61. ________, The conjugate gradient method for linear and nonlinear operator equations, SIAM J. Numer. Anal. 4 (1967), 10-26. MR 36 #1076.

62. W. Dotson, An application of ergodic theory to the solution of linear functional equations in Banach spaces, Bull. Amer. Math. Soc. 75 (1969), 347-352. MR 39 #795.

63. ________, On the Mann iterative process, Trans. Amer. Math. Soc. 149 (1970), 65-73. MR 41 #2477.

64. ________, On the solution of linear functional equations by averaging iteration, Proc. Amer. Math. Soc. 25 (1970), 504-506. MR 41 #4333.

65. ________, Mean ergodic theorems and iterative solution of linear functional equations, J. Math. Anal. Appl. 34 (1971), 141-150.

66. M. Edelstein, A remark on a theorem of M. A. Krasnoselski, Amer. Math. Monthly 73 (1966), 509-510. MR 33 #3072.

67. A. Esajan and V. Stetsenko, On the convergence of successive approximations for equations of the second kind, Dokl. Akad. Nauk Tadzik. SSR 7 (1964), no. 2, 13-35. (Russian)

68. D. de Figueiredo and L. Karlovitz, On the approximate solution of linear functional equations in Banach spaces, J. Math. Anal. Appl. 24 (1968), 654-664. MR 38 #6415.

69. V. Fridman, The convergence of the method of steepest descent, Uspehi Mat. Nauk 17 (1962), no. 3, 201-204. (Russian) MR 25 #3602.

70. C. Groetsch, Ergodic theory and iterative solution of linear operator equations, to appear, Applicable Analysis, see also Notices Amer. Math. Soc. 19 (1972), A-821.

71. R. Hayes, Iterative methods of solving linear problems on Hilbert space, National Bureau of Standards Applied Math. Ser. 39 (1954), 71-103. MR 16, 597.

72. W. Kammerer and M. Nashed, Steepest descent for singular linear operators with nonclosed range, Applicable Analysis 1 (1971), 143-159. MR 44 #7319.

73. ________, Iterative methods for best approximate solutions of linear integral equations of the first and second kinds, MRC Technical Summary Report #1117, Mathematics Research Center, University of Wisconsin, 1971.

74. ________, On the convergence of the conjugate gradient method for singular linear operator equations, SIAM J. Numer. Anal. 9 (1972), 165-181.

75. S. Kaniel, Construction of a fixed point for contractions in Banach space, Israel J. Math. 9 (1971), 535-540. MR 44 #2110.

76. L. Kantorovic, On the method of steepest descent, Dokl. Akad. Nauk SSSR 56 (1947), 233-236. (Russian) MR 9, 308.

77. ________, Functional Analysis and Applied Mathematics, National Bureau of Standards Report 1509, 1952. MR 10, 380. MR 14, 766.

78. ________, Approximate solution of functional equations, Uspehi Mat. Nauk 11 (1956), no. 6, 99-116. (Russian) MR 18, 747.

79. L. Kantorovic and G. Akilov, Functional Analysis in Normed Spaces, Pergamon Press, Oxford, 1964. MR 22 #9837.

80. L. Kantorovic and V. Krylov, Approximate Methods of Higher Analysis, Interscience, New York, 1964. MR 13, 77. MR 21 #5268.

81. B. Kellogg, An alternating direction method for operator equations, SIAM J. Appl. Math. 12 (1964), 848-854. MR 31 #4194.

82. L. Kivistik, On certain iteration methods for solving operator equations in Hilbert space, Eesti NSV Tead. Akad. Toimetised Füüs.-Mat. 9 (1960), 229-241. (Russian) MR 24 #A1035.

83. J. Koliha, On the iterative solution of linear operator equations with self-adjoint operators, J. Austral. Math. Soc. 13 (1972), 241-255.

84. ________, Convergent and stable operators and their generalization, to appear.

85. ________, Ergodic theory and averaging iterations, to appear.

86. J. Kolomý, On the solution of linear functional equations in Hilbert space, Casopis. Pest. Mat. 86 (1961), 314-317. MR 24 #A3518.

87. ________, On the solution of homogeneous functional equations in Hilbert space, Comment. Math. Univ. Carolinae 3 (1962), no. 4, 36-47. MR 26 #6796.

88. ________, On the solution of functional equations with linear bounded operators, Comment. Math. Univ. Carolinae 6 (1965), 141-143. MR 30 #5192.

89. ________, New methods for the solution of linear functional equations, Apl. Mat. 10 (1965), 246-248. (Russian) MR 31 #5307.

90. L. Korkina, The solution of operator equations of the first kind in Hilbert spaces, Izv. Vyss. Ucebn. Zaved. Mathematika 62 (1967), no. 7, 65-69. (Russian) MR 35 #7177.

91. V. Kostarcuk and B. Pugacev, Exact estimation of decrease of error in one step of the method of quickest descent, Voronez. Gos. Univ. Trudy Sem. Funkcional. Anal., no. 2 (1956), 25-30. (Russian) MR 18, 713.

92. M. Krasnoselskii, Two remarks on the method of successive approximations, Uspehi Mat. Nauk 10 (1955), no. 1, 123-127. (Russian) MR 16, 833.

93. ________, On the solution of equations with self-adjoint operators by the method of successive approximations, Uspehi Mat. Nauk 15 (1960), 161-165. (Russian) MR 22 #9866.

94. M. Krasnoselskii and V. Stecenko, Some notes on Seidel's method, USSR Comput. Math. and Math. Phys. 9 (1969), no. 1, 230-238. MR 40 #3332.

95. M. Krasnoselskii et al., Approximate Solutions of Operator Equations, Izdat. "Nauka", Moscow, 1969. (Russian) MR 41 #4271.

96. S. Krein and O. Prozorovskaja, An analogue of Seidel's method for operator equations, Voronez. Gos. Univ. Trudy Sem. Funkcional. Anal. 5 (1957), 35-38. (Russian) MR 20 #2081.

97. M. Kurpel, An approximate method of solving linear operator equations in Hilbert space, Dopovidi Akad. Nauk Ukrain RSR 1963, 1275-1279. (Ukrainian) MR 29 #2652.

98. ________, Conditions for convergence and error estimates for a general iterative method of solution of linear operator equations, First Republ. Math. Conf. of Young Researchers, Part I, 418-427, Akad. Nauk Ukrain. SSR Inst. Mat., Kiev, 1965. (Russian) MR 33 #4731.

99. N. Kurpel, Projection-iterative Methods of Solution of Operator Equations, Izdat. "Naukova Dumka", Kiev, 1968. (Russian) MR 40 #7898.

100. Y. Kwon and R. Redheffer, Remarks on linear equations in Banach space, Arch. Rational Mech. Anal. 32 (1969), 247-254. MR 40 #6284.

101. V. Lebedev, Iterative methods for solving operator equations with a spectrum contained in several intervals, USSR Comput. Math. and Math. Phys. 9 (1969), no. 6, 17-24. MR 42 #7052.

102. V. Lebedev and S. Finogenov, The order of choice of the iteration parameters in the cyclic Cebysev iteration method, Z. Vycisl. Mat. i Mat. Fiz 11 (1971), 425-438. (Russian) MR 44 #3479.

103. W. Lipfert, Iterative Lösung der linearen Gleichung Ax=y unter Verwendung einer Näherung für den inversen Operator von A, Wiss. Z. Techn. Univ. Dresden 19 (1970), 399-404. MR 43 #3816.

104. A. Lucka, The Method of Averaging Functional Corrections, Academic Press, New York, 1965. MR 30 #4394. MR 32 #3298.

105. W. Mann, Mean value methods in iteration, Proc. Amer. Math. Soc. 4 (1953), 506-510. MR 14, 988.

106. I. Marek. On the SOR method for solving linear equations in Banach spaces, Wiss. Z. Techn. Hochsch. Karl-Marx-Stadt 11 (1969), 335-341. MR 43 #3818.

107. A. Martynjuk, Some new criteria for convergence of the method of successive approximations, Kazan. Gos. Univ. Ucen. Zap. 124 (1964), no. 6, 183-188. (Russian) MR 32 #4573.

108. S. Mihlin, Variational Methods in Mathematical Physics, Pergamon Press, Macmillan, New York, 1964. MR 22 #1981. MR 30 #2712.

109. B. Mosolov, Necessary and sufficient conditions of the solvability of operator equations of the first kind, Izv. Akad. Nauk UzSSR Ser. Fiz.-Mat. Nauk 13 (1969), no. 3, 78-79. (Russian) MR 43 #5329.

110. M. Nashed, Steepest descent for singular linear operator equations, SIAM J. Numer. Anal. 7 (1970), 358-362. MR 42 #3989.

111. ________, Generalized inverses, normal solvability, and iteration for singular operator equations, Proc. Adv. Seminar Math. Res. Cent., U. of Wisc. 26 (1970), 311-359, Academic Press, New York, 1971. MR 43 #1003.

112. W. Niethammer and W. Schempp, On the construction of iteration methods for linear equations in Banach spaces by summation methods, Aequationes Math. 5 (1970), 273-284. MR 44 #7695.

113. L. Oblomskaja, Methods of successive approximation for linear equations in Banach spaces, USSR Compt. Math. and Math. Phys. 8 (1968), no. 2, 239-253. MR 38 #587.

114. C. Outlaw and C. Groetsch, Averaging iteration in a Banach space, Bull. Amer. Math. Soc. 75 (1969), 430-432. MR 39 #835.

115. L. Peklova, The gradient version of Ju. D. Sokolov's method, Ukrain. Mat. Z. 21 (1969), 712-713. (Russian) MR 41 #4793.

116. W. Petryshyn, Direct and iterative methods for the solution of linear operator equations in Hilbert space, Trans. Amer. Math. Soc. 105 (1962), 136-175. MR 26 #3180.

117. ________, The generalized overrelaxation method for the approximate solution of operator equations in Hilbert space, SIAM J. Appl. Math. 10 (1962), 675-690. MR 29 #3003.

118. ________, On a general iterative method for the approximate solution of linear operator equations, Math. Comp. 17 (1963), 1-10. MR 29 #729.

119. ________, On the generalized overrelaxation method for operation equations, Proc. Amer. Math. Soc. 14 (1963), 917-924. MR 29 #6652.

120. ________, On the extrapolated Jacobi or simultaneous displacements method in the solution of matrix and operator equations, Math. Comp. 19 (1965), 37-55. MR 31 #873.

121. ________, On two variants of a method for the solution of linear equations with unbounded operators and their applications, J. Math. and Phys. 44 (1965), 297-312. MR 33 #7902.

122. ________, On a class of K-p.d. and non K-p.d. operators and operator equations, J. Math. Anal. Appl. 10 (1965), 1-24. MR 35 #7141.

123. ________, On the convergence of an accelerated iterative method in the solution of singular equations, ICR Quarterly Report, no. 9, Univ. of Chicago, 1966.

124. ________, On generalized inverses and uniform convergence of $(I-\beta K)^n$ with applications to iterative methods, J. Math. Anal. Appl. 18 (1967), 417-439. MR 34 #8191.

125. ________, On projectional solvability and the Fredholm alternative for equations involving linear A-proper operators, Arch. Rational Mech. Anal. 30 (1968), 270-284. MR 37 #6776.

126. ________, Remarks on generalized overrelaxation and the extrapolated Jacobi methods for operator equations in Hilbert space, J. Math. Anal. Appl. 29 (1970), 558-568. MR 41 #7458.

127. J. de Pillis, Gauss-Seidel convergence for operators on Hilbert space, SIAM J. Numer. Anal. 10 (1973), 112-122.

128. L. Rall, Error bounds for iterative solutions of Fredholm integral equations, Pacific J. Math. 5 (1955), 977-986. MR 18, 72.

129. ________, Computational Solutions of Nonlinear Operator Equations, John Wiley and Sons, New York, 1969. MR 39 #2289.

130. J. Reinermann, Über Fixpunkte kontrahierender Abbildungen und schwach konvergente Toeplitz-Verfahren, Arch. Math. (Basel) 20 (1969), 59-64. MR 43 #2583.

131. ________, Über Toeplitzsche Iterationsverfahren und einige ihrer Anwendungen in der konstruktiven Fixpunktheorie, Studia Math. 32 (1969), 209-227.

132. P. Rosenbloom, The method of steepest descent, Proc. of Symposia in Appl. Math. 6 (1956), 127-176. MR 18, 71.

133. T. Sabirov and A. Esajan, The question of convergence of the averaging method of functional corrections, Dokl. Akad. Nauk Tadzik. SSR 9 (1966), no. 1, 8-12. (Russian) MR 34 #1893.

134. V. Samanskii, Convergence of iterative processes, Ukrain. Mat. Z. 13 (1961), 113-115. MR 26 #1989.

135. ________, On some computational schemes for iterative processes, Ukrain. Mat. Z. 14 (1962), 100-109. (Russian) MR 26 #4464.

136. B. Samokis, Investigation of the rapidity of convergence of the method of steepest descent, Uspehi Mat. Nauk 12 (1957), no. 1, 238-240. (Russian) MR 19, 322.

137. H. Schaefer, Über die Methode sukzessiver Approximationen, Jber. Deutsch. Math.-Verein. 59 (1957), 131-140. MR 18, 811.

138. W. Schempp, Iterative solution of linear operator equations in Hilbert space and optimal Euler methods, Arch. Math. (Basel) 21 (1970), 390-395. MR 43 #2539.

139. M. Schönberg, Sur la méthode d'itération de Wiarda et Bückner pour la résolution de l'équation de Fredholm I, Acad. Roy. Belg. Bull. Cl. Sci. 37 (1951), 1141-1156. MR 13, 952.

140. ________, Sur la méthode d'itération de Wiarda et Bückner pour la résolution de l'équation de Fredholm II, Acad. Roy. Belg. Bull. Cl. Sci. 38 (1952), 154-167. MR 13, 952.

141. P. Sobolevskii, On equations with operators forming an acute angle, Dokl. Akad. Nauk SSSR 116 (1957), 754-757. (Russian) MR 20 #4194.

142. Ju. Sokolov, Sur la méthode du moyennage des corrections fonctionnelles, Ukrain. Mat. Z. 9 (1957), 82-100. (Russian) MR 19, 687.

143. ________, The Method of Averaging Functional Corrections, Izdat. "Naukova Dumka", Kiev, 1967. (Russian) MR 35 #3931.

144. W. Strang, On the Kantorovich inequality, Proc. Amer. Math. Soc. 11 (1960), 468. MR 22 #2904.

145. J. Wessinger, Verallgemeinerungen des Seidelschen Iterationsverfahrens, Z. Angew. Math. Mech. 33 (1953), 155-163. MR 15, 66.

Vol. 215: P. Antonelli, D. Burghelea and P. J. Kahn, The Concordance-Homotopy Groups of Geometric Automorphism Groups. X, 140 pages. 1971. DM 16,-

Vol. 216: H. Maaß, Siegel's Modular Forms and Dirichlet Series. VII, 328 pages. 1971. DM 20,-

Vol. 217: T. J. Jech, Lectures in Set Theory with Particular Emphasis on the Method of Forcing. V, 137 pages. 1971. DM 16,-

Vol. 218: C. P. Schnorr, Zufälligkeit und Wahrscheinlichkeit. IV, 212 Seiten. 1971. DM 20,-

Vol. 219: N. L. Alling and N. Greenleaf, Foundations of the Theory of Klein Surfaces. IX, 117 pages. 1971. DM 16,-

Vol. 220: W. A. Coppel, Disconjugacy. V, 148 pages. 1971. DM 16,-

Vol. 221: P. Gabriel und F. Ulmer, Lokal präsentierbare Kategorien. V, 200 Seiten. 1971. DM 18,-

Vol. 222: C. Meghea, Compactification des Espaces Harmoniques. III, 108 pages. 1971. DM 16,-

Vol. 223: U. Felgner, Models of ZF-Set Theory. VI, 173 pages. 1971. DM 16,-

Vol. 224: Revètements Etales et Groupe Fondamental. (SGA 1). Dirigé par A. Grothendieck XXII, 447 pages. 1971. DM 30,-

Vol. 225: Théorie des Intersections et Théorème de Riemann-Roch. (SGA 6). Dirigé par P. Berthelot, A. Grothendieck et L. Illusie. XII, 700 pages. 1971. DM 40,-

Vol. 226: Seminar on Potential Theory, II. Edited by H. Bauer. IV, 170 pages. 1971. DM 18,-

Vol. 227: H. L. Montgomery, Topics in Multiplicative Number Theory. IX, 178 pages. 1971. DM 18,-

Vol. 228: Conference on Applications of Numerical Analysis. Edited by J. Ll. Morris. X, 358 pages. 1971. DM 26,-

Vol. 229: J. Väisälä, Lectures on n-Dimensional Quasiconformal Mappings. XIV, 144 pages. 1971. DM 16,-

Vol. 230: L. Waelbroeck, Topological Vector Spaces and Algebras. VII, 158 pages. 1971. DM 16,-

Vol. 231: H. Reiter, L¹-Algebras and Segal Algebras. XI, 113 pages. 1971. DM 16,-

Vol. 232: T. H. Ganelius, Tauberian Remainder Theorems. VI, 75 pages. 1971. DM 16,-

Vol. 233: C. P. Tsokos and W. J. Padgett. Random Integral Equations with Applications to stochastic Systems. VII, 174 pages. 1971. DM 18,-

Vol. 234: A. Andreotti and W. Stoll. Analytic and Algebraic Dependence of Meromorphic Functions. III, 390 pages. 1971. DM 26,-

Vol. 235: Global Differentiable Dynamics. Edited by O. Hájek, A. J. Lohwater, and R. McCann. X, 140 pages. 1971. DM 16,-

Vol. 236: M. Barr, P. A. Grillet, and D. H. van Osdol. Exact Categories and Categories of Sheaves. VII, 239 pages. 1971. DM 20,-

Vol. 237: B. Stenström, Rings and Modules of Quotients. VII, 136 pages. 1971. DM 16,-

Vol. 238: Der kanonische Modul eines Cohen-Macaulay-Rings. Herausgegeben von Jürgen Herzog und Ernst Kunz. VI, 103 Seiten. 1971. DM 16,-

Vol. 239: L. Illusie, Complexe Cotangent et Déformations I. XV, 355 pages. 1971. DM 26,-

Vol. 240: A. Kerber, Representations of Permutation Groups I. VII, 192 pages. 1971. DM 18,-

Vol. 241: S. Kaneyuki, Homogeneous Bounded Domains and Siegel Domains. V, 89 pages. 1971. DM 16,-

Vol. 242: R. R. Coifman et G. Weiss, Analyse Harmonique Non-Commutative sur Certains Espaces. V, 160 pages. 1971. DM 16,-

Vol. 243: Japan-United States Seminar on Ordinary Differential and Functional Equations. Edited by M. Urabe. VIII, 332 pages. 1971. DM 26,-

Vol. 244: Séminaire Bourbaki - vol. 1970/71. Exposés 382-399. IV, 356 pages. 1971. DM 26,-

Vol. 245: D. E. Cohen, Groups of Cohomological Dimension One. V, 99 pages. 1972. DM 16,-

Vol. 246: Lectures on Rings and Modules. Tulane University Ring and Operator Theory Year, 1970-1971. Volume I. X, 661 pages. 1972. DM 40,-

Vol. 247: Lectures on Operator Algebras. Tulane University Ring and Operator Theory Year, 1970-1971. Volume II. XI, 786 pages. 1972. DM 40,-

Vol. 248: Lectures on the Applications of Sheaves to Ring Theory. Tulane University Ring and Operator Theory Year, 1970-1971. Volume III. VIII, 315 pages. 1971. DM 26,-

Vol. 249: Symposium on Algebraic Topology. Edited by P. J. Hilton. VII, 111 pages. 1971. DM 16,-

Vol. 250: B. Jónsson, Topics in Universal Algebra. VI, 220 pages. 1972. DM 20,-

Vol. 251: The Theory of Arithmetic Functions. Edited by A. A. Gioia and D. L. Goldsmith VI, 287 pages. 1972. DM 24,-

Vol. 252: D. A. Stone, Stratified Polyhedra. IX, 193 pages. 1972. DM 18,-

Vol. 253: V. Komkov, Optimal Control Theory for the Damping of Vibrations of Simple Elastic Systems. V, 240 pages. 1972. DM 20,-

Vol. 254: C. U. Jensen, Les Foncteurs Dérivés de $\varprojlim$ et leurs Applications en Théorie des Modules. V, 103 pages. 1972. DM 16,-

Vol. 255: Conference in Mathematical Logic - London '70. Edited by W. Hodges. VIII, 351 pages. 1972. DM 26,-

Vol. 256: C. A. Berenstein and M. A. Dostal, Analytically Uniform Spaces and their Applications to Convolution Equations. VII, 130 pages. 1972. DM 16,-

Vol. 257: R. B. Holmes, A Course on Optimization and Best Approximation. VIII, 233 pages. 1972. DM 20,-

Vol. 258: Séminaire de Probabilités VI. Edited by P. A. Meyer. VI, 253 pages. 1972. DM 22,-

Vol. 259: N. Moulis, Structures de Fredholm sur les Variétés Hilbertiennes. V, 123 pages. 1972. DM 16,-

Vol. 260: R. Godement and H. Jacquet, Zeta Functions of Simple Algebras. IX, 188 pages. 1972. DM 18,-

Vol. 261: A. Guichardet, Symmetric Hilbert Spaces and Related Topics. V, 197 pages. 1972. DM 18,-

Vol. 262: H. G. Zimmer, Computational Problems, Methods, and Results in Algebraic Number Theory. V, 103 pages. 1972. DM 16,-

Vol. 263: T. Parthasarathy, Selection Theorems and their Applications. VII, 101 pages. 1972. DM 16,-

Vol. 264: W. Messing, The Crystals Associated to Barsotti-Tate Groups: With Applications to Abelian Schemes. III, 190 pages. 1972. DM 18,-

Vol. 265: N. Saavedra Rivano, Catégories Tannakiennes. II, 418 pages. 1972. DM 26,-

Vol. 266: Conference on Harmonic Analysis. Edited by D. Gulick and R. L. Lipsman. VI, 323 pages. 1972. DM 24,-

Vol. 267: Numerische Lösung nichtlinearer partieller Differential- und Integro-Differentialgleichungen. Herausgegeben von R. Ansorge und W. Törnig, VI, 339 Seiten. 1972. DM 26,-

Vol. 268: C. G. Simader, On Dirichlet's Boundary Value Problem. IV, 238 pages. 1972. DM 20,-

Vol. 269: Théorie des Topos et Cohomologie Etale des Schémas. (SGA 4). Dirigé par M. Artin, A. Grothendieck et J. L. Verdier. XIX, 525 pages. 1972. DM 50,-

Vol. 270: Théorie des Topos et Cohomologie Etale des Schémas. Tome 2. (SGA 4). Dirigé par M. Artin, A. Grothendieck et J. L. Verdier. V, 418 pages. 1972. DM 50,-

Vol. 271: J. P. May, The Geometry of Iterated Loop Spaces. IX, 175 pages. 1972. DM 18,-

Vol. 272: K. R. Parthasarathy and K. Schmidt, Positive Definite Kernels, Continuous Tensor Products, and Central Limit Theorems of Probability Theory. VI, 107 pages. 1972. DM 16,-

Vol. 273: U. Seip, Kompakt erzeugte Vektorräume und Analysis. IX, 119 Seiten. 1972. DM 16,-

Vol. 274: Toposes, Algebraic Geometry and Logic. Edited by. F. W. Lawvere. VI, 189 pages. 1972. DM 18,-

Vol. 275: Séminaire Pierre Lelong (Analyse) Année 1970-1971. VI, 181 pages. 1972. DM 18,-

Vol. 276: A. Borel, Représentations de Groupes Localement Compacts. V, 98 pages. 1972. DM 16,-

Vol. 277: Séminaire Banach. Edité par C. Houzel. VII, 229 pages. 1972. DM 20,-

Vol. 278: H. Jacquet, Automorphic Forms on GL(2). Part II. XIII, 142 pages. 1972. DM 16,–

Vol. 279: R. Bott, S. Gitler and I. M. James, Lectures on Algebraic and Differential Topology. V, 174 pages. 1972. DM 18,–

Vol. 280: Conference on the Theory of Ordinary and Partial Differential Equations. Edited by W. N. Everitt and B. D. Sleeman. XV, 367 pages. 1972. DM 26,–

Vol. 281: Coherence in Categories. Edited by S. Mac Lane. VII, 235 pages. 1972. DM 20,–

Vol. 282: W. Klingenberg und P. Flaschel, Riemannsche Hilbertmannigfaltigkeiten. Periodische Geodätische. VII, 211 Seiten. 1972. DM 20,–

Vol. 283: L. Illusie, Complexe Cotangent et Déformations II. VII, 304 pages. 1972. DM 24,–

Vol. 284: P. A. Meyer, Martingales and Stochastic Integrals I. VI, 89 pages. 1972. DM 16,–

Vol. 285: P. de la Harpe, Classical Banach-Lie Algebras and Banach-Lie Groups of Operators in Hilbert Space. III, 160 pages. 1972. DM 16,–

Vol. 286: S. Murakami, On Automorphisms of Siegel Domains. V, 95 pages. 1972. DM 16,–

Vol. 287: Hyperfunctions and Pseudo-Differential Equations. Edited by H. Komatsu. VII, 529 pages. 1973. DM 36,–

Vol. 288: Groupes de Monodromie en Géométrie Algébrique. (SGA 7 I). Dirigé par A. Grothendieck. IX, 523 pages. 1972. DM 50,–

Vol. 289: B. Fuglede, Finely Harmonic Functions. III, 188. 1972. DM 18,–

Vol. 290: D. B. Zagier, Equivariant Pontrjagin Classes and Applications to Orbit Spaces. IX, 130 pages. 1972. DM 16,–

Vol. 291: P. Orlik, Seifert Manifolds. VIII, 155 pages. 1972. DM 16,–

Vol. 292: W. D. Wallis, A. P. Street and J. S. Wallis, Combinatorics: Room Squares, Sum-Free Sets, Hadamard Matrices. V, 508 pages. 1972. DM 50,–

Vol. 293: R. A. DeVore, The Approximation of Continuous Functions by Positive Linear Operators. VIII, 289 pages. 1972. DM 24,–

Vol. 294: Stability of Stochastic Dynamical Systems. Edited by R. F. Curtain. IX, 332 pages. 1972. DM 26,–

Vol. 295: C. Dellacherie, Ensembles Analytiques, Capacités, Mesures de Hausdorff. XII, 123 pages. 1972. DM 16,–

Vol. 296: Probability and Information Theory II. Edited by M. Behara, K. Krickeberg and J. Wolfowitz. V, 223 pages. 1973. DM 20,–

Vol. 297: J. Garnett, Analytic Capacity and Measure. IV, 138 pages. 1972. DM 16,–

Vol. 298: Proceedings of the Second Conference on Compact Transformation Groups. Part 1. XIII, 453 pages. 1972. DM 32,–

Vol. 299: Proceedings of the Second Conference on Compact Transformation Groups. Part 2. XIV, 327 pages. 1972. DM 26,–

Vol. 300: P. Eymard, Moyennes Invariantes et Représentations Unitaires. II, 113 pages. 1972. DM 16,–

Vol. 301: F. Pittnauer, Vorlesungen über asymptotische Reihen. VI, 186 Seiten. 1972. DM 18,–

Vol. 302: M. Demazure, Lectures on p-Divisible Groups. V, 98 pages. 1972. DM 16,–

Vol. 303: Graph Theory and Applications. Edited by Y. Alavi, D. R. Lick and A. T. White. IX, 329 pages. 1972. DM 26,–

Vol. 304: A. K. Bousfield and D. M. Kan, Homotopy Limits, Completions and Localizations. V, 348 pages. 1972. DM 26,–

Vol. 305: Théorie des Topos et Cohomologie Etale des Schémas. Tome 3. (SGA 4). Dirigé par M. Artin, A. Grothendieck et J. L. Verdier. VI, 640 pages. 1973. DM 50,–

Vol. 306: H. Luckhardt, Extensional Gödel Functional Interpretation. VI, 161 pages. 1973. DM 18,–

Vol. 307: J. L. Bretagnolle, S. D. Chatterji et P.-A. Meyer, Ecole d'été de Probabilités: Processus Stochastiques. VI, 198 pages. 1973. DM 20,–

Vol. 308: D. Knutson, λ-Rings and the Representation Theory of the Symmetric Group. IV, 203 pages. 1973. DM 20,–

Vol. 309: D. H. Sattinger, Topics in Stability and Bifurcation Theory. VI, 190 pages. 1973. DM 18,–

Vol. 310: B. Iversen, Generic Local Structure of the Morphisms in Commutative Algebra. IV, 108 pages. 1973. DM 16,–

Vol. 311: Conference on Commutative Algebra. Edited by J. W. Brewer and E. A. Rutter. VII, 251 pages. 1973. DM 22,–

Vol. 312: Symposium on Ordinary Differential Equations. Edited by W. A. Harris, Jr. and Y. Sibuya. VIII, 204 pages. 1973. DM 22,–

Vol. 313: K. Jörgens and J. Weidmann, Spectral Properties of Hamiltonian Operators. III, 140 pages. 1973. DM 16,–

Vol. 314: M. Deuring, Lectures on the Theory of Algebraic Functions of One Variable. VI, 151 pages. 1973. DM 16,–

Vol. 315: K. Bichteler, Integration Theory (with Special Attention to Vector Measures). VI, 357 pages. 1973. DM 26,–

Vol. 316: Symposium on Non-Well-Posed Problems and Logarithmic Convexity. Edited by R. J. Knops. V, 176 pages. 1973. DM 18,–

Vol. 317: Séminaire Bourbaki – vol. 1971/72. Exposés 400–417. IV, 361 pages. 1973. DM 26,–

Vol. 318: Recent Advances in Topological Dynamics. Edited by A. Beck, VIII, 285 pages. 1973. DM 24,–

Vol. 319: Conference on Group Theory. Edited by R. W. Gatterdam and K. W. Weston. V, 188 pages. 1973. DM 18,–

Vol. 320: Modular Functions of One Variable I. Edited by W. Kuyk. V, 195 pages. 1973. DM 18,–

Vol. 321: Séminaire de Probabilités VII. Edité par P. A. Meyer. VI, 322 pages. 1973. DM 26,–

Vol. 322: Nonlinear Problems in the Physical Sciences and Biology. Edited by I. Stakgold, D. D. Joseph and D. H. Sattinger. VIII, 357 pages. 1973. DM 26,–

Vol. 323: J. L. Lions, Perturbations Singulières dans les Problèmes aux Limites et en Contrôle Optimal. XII, 645 pages. 1973. DM 42,–

Vol. 324: K. Kreith, Oscillation Theory. VI, 109 pages. 1973. DM 16,–

Vol. 325: Ch.-Ch. Chou, La Transformation de Fourier Complexe et L'Equation de Convolution. IX, 137 pages. 1973. DM 16,–

Vol. 326: A. Robert, Elliptic Curves. VIII, 264 pages. 1973. DM 22,–

Vol. 327: E. Matlis, 1-Dimensional Cohen-Macaulay Rings. XII, 157 pages. 1973. DM 18,–

Vol. 328: J. R. Büchi and D. Siefkes, The Monadic Second Order Theory of All Countable Ordinals. VI, 217 pages. 1973. DM 20,–

Vol. 329: W. Trebels, Multipliers for (C, α)-Bounded Fourier Expansions in Banach Spaces and Approximation Theory. VII, 103 pages. 1973. DM 16,–

Vol. 330: Proceedings of the Second Japan-USSR Symposium on Probability Theory. Edited by G. Maruyama and Yu. V. Prokhorov. VI, 550 pages. 1973. DM 36,–

Vol. 331: Summer School on Topological Vector Spaces. Edited by L. Waelbroeck. VI, 226 pages. 1973. DM 20,–

Vol. 332: Séminaire Pierre Lelong (Analyse) Année 1971-1972. V, 131 pages. 1973. DM 16,–

Vol. 333: Numerische, insbesondere approximationstheoretische Behandlung von Funktionalgleichungen. Herausgegeben von R. Ansorge und W. Törnig. VI, 296 Seiten. 1973. DM 24,–

Vol. 334: F. Schweiger, The Metrical Theory of Jacobi-Perron Algorithm. V, 111 pages. 1973. DM 16,–

Vol. 335: H. Huck, R. Roitzsch, U. Simon, W. Vortisch, R. Walden, B. Wegner und W. Wendland, Beweismethoden der Differentialgeometrie im Großen. IX, 159 Seiten. 1973. DM 18,–

Vol. 336: L'Analyse Harmonique dans le Domaine Complexe. Edité par E. J. Akutowicz. VIII, 169 pages. 1973. DM 18,–

Vol. 337: Cambridge Summer School in Mathematical Logic. Edited by A. R. D. Mathias and H. Rogers. IX, 660 pages. 1973. DM 42,–

Vol. 338: J. Lindenstrauss and L. Tzafriri, Classical Banach Spaces. IX, 243 pages. 1973. DM 22,–

Vol. 339: G. Kempf, F. Knudsen, D. Mumford and B. Saint-Donat, Toroidal Embeddings I. VIII, 209 pages. 1973. DM 20,–

Vol. 340: Groupes de Monodromie en Géométrie Algébrique. (SGA 7 II). Par P. Deligne et N. Katz. X, 438 pages. 1973. DM 40,–

Vol. 341: Algebraic K-Theory I, Higher K-Theories. Edited by H. Bass. XV, 335 pages. 1973. DM 26,–

Vol. 342: Algebraic K-Theory II, "Classical" Algebraic K-Theory, and Connections with Arithmetic. Edited by H. Bass. XV, 527 pages. 1973. DM 36,–